RELUCTANT PARTNERS?

Insufficiencies and inequities in food production and supply in poor countries need to be addressed as problems of both agricultural resource management and rural democratization.

Reluctant Partners? combines comprehensive empirical insights into NGOs' work in agriculture with wider considerations of their relations with the state and their contribution to democratic pluralism.

This overview volume for the *Non-Governmental Organizations* series contextualizes and synthesizes the case study material in the three regional volumes on *Africa, Asia* and *Latin America,* where over sixty specially commissioned case studies of farmer participatory approaches to agricultural innovation are presented.

Specific questions are raised. How good/bad are NGOs at promoting technological innovation and addressing constraints to change in peasant agriculture? How effective are NGOs at strengthening grassroots/local organizations? How do/will donor pressures influence NGOs and their links to the state?

John Farrington and **Anthony Bebbington** are Research Fellows with, and **Kate Wellard** and **David J. Lewis** are associates of the Overseas Development Institute.

NON-GOVERNMENTAL ORGANIZATIONS SERIES
Co-ordinated by the Overseas Development Institute

The titles in the series are:

RELUCTANT PARTNERS?

Non-governmental organizations,
the state and sustainable
agricultural development

John Farrington
and
Anthony Bebbington
with
Kate Wellard
and
David J. Lewis

London and New York

First published 1993
by Routledge
11 New Fetter Lane

Simultaneously published in the USA and Canada
by Routledge
29 West 35th Street, New York, NY 10001

© 1993 Overseas Development Institute

Typeset by Witwell Ltd, Southport, in 10/12 point Garamond
Printed and bound in Great Britain by
Biddles Ltd, Guildford and Kings' Lynn

British Library Cataloguing in Publication Data
A catalogue reference for this book is available from the British
Library.

Library of Congress Cataloging in Publication Data has been applied
for.

ISBN 0–415–08843–7   Hbk
ISBN 0–415–08844–5   Pbk

CONTENTS

FIGURES

TABLES

BOXES

FOREWORD

Non-governmental organizations of various types and sizes emerging throughout the Third World now constitute a promising but not yet firmly established 'third sector'. This sector is different from but interacts with both the public (state) and private (for-profit) sectors. It generally complements but often challenges them, as chronicled and analysed by Carroll (1992), Clark (1991), Edwards and Hulme (1992a), Fisher-Peck (1993), and Korten (1990), among others. This examination of NGOs by John Farrington and colleagues at the Overseas Development Institute builds on three years of work and some seventy case studies. It sheds much light on NGOs' potentials and their limitations in the agricultural sector. Since most of the world's poor continue to depend on agriculture for their sustenance, this examination is especially welcome.

Over the last decade, the NGO movement has matured and gained both momentum and support. It is now not so much endangered by opposition from official or commercial sources as by a variety of embraces. Donor agencies and governments increasingly seek to 'utilize' NGOs, in the process promoting modes and scales of operation that can jeopardize NGO performance and integrity. Burgeoning opportunities for NGOs often attract some unscrupulous or sloppy operators. Enthusiasts who are excited by the political and ideological possibilities of NGOs creating an alternative development can become oblivious or apologetic about shortcomings. The kind of broad-ranging and empirical assessment of NGOs' actual performance, for better and for worse, which ODI has undertaken here is thus timely as well as welcome.

Governments appear more and more open to collaborative relationships with NGOs. Last April, while in Ghana, I met a newly appointed deputy director in the Ministry of Agriculture who was assigned specifically to work as liaison with NGOs engaged in agricultural development in that country. When in Indonesia a few months later, the head of one of that country's most praiseworthy NGOs told me how he and thirty other NGO representatives had recently been invited to a meeting with the Minister of Agriculture, who stated his ministry's intention to work more closely with NGOs across a broad range of activities. During the same visit, a top official in Indonesia's planning ministry described plans for implementing rural infrastructure projects worth US$200 million through NGOs to ensure that benefits reach the poor. Such embraces can, indeed, be more a cause for concern than for celebration.

When such large amounts of money are involved, the potential for disappointment and even disaster with NGOs looms large. A highly respected NGO leader in India when I asked him how many NGOs in his country were both 'genuine' and effective, sadly guessed this would be around 20 per cent. Both of us regretted that proponents of the NGO alternative were not more critical of the limitations and even deceptions one could find on close examination of this heterogeneous category of organizations.

Farrington and his colleagues appreciate both the variety of NGOs and the need for objective evaluation. Their focus on the subset of NGOs classified as service organizations, which work with grassroots organizations to motivate and improve their performance, is well-chosen (Carroll 1992). By inviting researchers and practitioners from African, Asian and Latin American countries to furnish documentation on NGO experience in the agricultural sector, they have mobilized enough empirical material to fill three companion volumes assessing the contributions of NGOs across much of the Third World, providing a data base and underpinning for their analysis here.

There is reason for enthusiasm and excitement about the prospects of 'third sector' development. We read here, for example, about how Bolivian officials, wishing to buffer the adverse effects on the poor of structural adjustment policies, were able to meet urgent human needs on a broad scale under difficult circumstances by supporting the work of large numbers of indigenous NGOs through a Social Emergency Fund (see Durán 1990). This helped to persuade the World Bank that it could indeed improve upon its *modus operandi* favouring large-scale government-managed projects in the social sector by 'retailing' opportunities through NGOs.

An experience, outside the geographical remit of the present study, which helped persuade USAID of the advantages attainable from working with NGOs was an agroforestry project in Haiti. Channelling technical and material assistance to groups of farmers through selected NGOs resulted in 75,000 peasants planting twenty million trees in four years, instead of 6,000 planting six million trees in five years as planned, and with a higher survival rate than expected. This success was due not just to the organizational strategy, however, but depended also on appropriate technology and incentives (Murray 1986).

While Bolivia under structural adjustment represented a difficult context for grassroots development, Haiti in the mid-1980s was an even more formidable challenge. The programmatic accomplishments of NGOs such as the Bangladesh Rural Advancement Committee (Lovell 1992) and Proshika (Wood and Palmer-Jones 1990) show what NGOs can do under adverse circumstances in still another part of the world.[1]

Many issues need to be resolved for NGOs to become still more widespread and effective in their provision of services and their energization of societies. It is particularly useful to read in this volume of cases where NGOs have undertaken agricultural research and extension, functions traditionally thought of as best suited for the public sector, because of their public-goods nature and the externalities involved. NGOs are also learning how to use market mechanisms to become more self-financing for services such as artificial insemination. So the great divide between public sector and private sector activities is being bridged by NGOs with feet on both sides.

The fundamental basis of the NGO sector is voluntarism, eschewing the resort to authority and coercive means exercised by the state and forswearing (sometimes not entirely) the profit incentives of private enterprises. While recognizing that self-interested motivation has an inescapable part in any endeavour, NGOs seek to tap the social energies which can come from positive-sum, other-oriented action.[2] This is something which stems from what economists would describe in technical terms as 'interdependence of utility functions'. In plain language, this means valuing others' well-being in addition to or as well as one's own.

Governments justify their existence by taking on tasks which are not well performed, or which are neglected, when individual self-interest governs behaviour. This makes NGOs competitors with the state within the domain of altruistic action, though ideally they ought to be allies. The potential for conflict becomes greater when either the state or NGOs are operating with less than exalted motives. Their relationship is nicely characterized in the title of this book: *Reluctant Partners?* There is as yet no final answer to the question, but it is one well worth exploring.

Norman Uphoff
Cornell University

NOTES

1 'Between 1985 and 1991, the percentage of infants and young children immunized against the six most common childhood diseases rose from 2 per cent to 62 per cent, and the figure is expected to reach 85 per cent within the next five years, according to World Bank figures. Non-governmental organizations have carried out the immunization program in neighborhood centers with the help of the US Agency for International Development and other donors.' *New York Times*, 19 March 1992, p.A8.
2 The important concept of 'social energy' has been introduced by Hirschman (1984). Its dynamics and sources are discussed analytically in Uphoff (1992: esp. 284–9, 352–6, 367–81).

PREFACE

REVOLUTIONS NEED INSTITUTIONS

At a workshop held in October 1992, Robert Chambers referred to the spread of farmer participatory methods in the development of agricultural technologies as a 'methodological revolution' in contemporary agricultural research (see also Pretty and Chambers 1992).

Revolutions, however, need institutions if their fruits are to be harvested sustainably. While the literature on methods for promoting farmer participation has burgeoned over that last decade, we lag far behind in understanding how that participation can be institutionalized. As long as these institutional questions remain unresolved, participation will continue to be an event rather than a process, and the prospects of constructing an agricultural development that is more equitable, democratic and appropriate for the rural poor will remain unfulfilled.

The Agricultural Research and Extension Network at ODI has itself contributed to the development and discussion of participatory methods through its programme of research and publication. As part of that research, it conducted a review of the largely unpublished literature on participatory methods for the 'Farmer First' conference in 1987. Much of the literature came from members of the Network. One outcome of that review was to draw attention to the fact that many of the institutions pioneering participatory approaches were non-governmental development organizations.

It rapidly became clear that many NGOs saw a role for themselves as a 'missing link' not simply to develop participatory methods, but, equally importantly, to empower the rural poor to contribute to technical change from their own resources, and to articulate demands on government services more effectively. Historically, the high levels of organization among farmers in the North has allowed them to influence policy – whether on agricultural technology or more widely – without the development of a distinct body of participatory methods, or of support organizations such as NGOs. Much the same can be said of the larger-scale farmers currently operating in many developing countries. By contrast, organizations of the rural poor remain weak. The potential of NGOs for strengthening the influence of these on agricultural change – a potential that might be enhanced by the increasing volume of resources at the command of NGOs – was a major reason for focusing on them in the present study.

The other source of inspiration for the study came from the feeling that any viable strategy of institutionalization would ultimately have to involve state institutions. Again, through the Network membership, we were aware of cases in which NGOs and government agricultural research institutes had been able to work together in such a way that the NGO became the vehicle through which farmer preferences and knowledge were fed back into research activities in the public sector. Although few in number, such cases pointed to wider opportunities, but also to the constraints and preconditions likely to be faced in efforts to bring the two sides together. They prompted a much wider search for evidence of what might be achieved, and how.

This book, and its companion volumes, report on the findings of this wider study. In searching for cases in which NGOs had engaged with government research institutes (and vice versa), we were interested in the range of motives underlying those relationships: NGOs entered into some relationships with a view to improving the quality of their own work; they entered others with the goal of exercising some form of pressure over the government institute in order to orient its work more directly toward the requirements of resource poor farmers. In further cases NGOs studiously avoided contact with government. An equally wide range of motives was found to underlie government organizations' efforts to work with – or ignore – NGOs.

In all, some seventy cases were documented, and were drawn from eighteen countries in Africa, Asia and Latin America. Most are written from the NGO perspective, but some from that of government, and in several cases we document the perspectives from each side on a particular set of experiences. Each paper discussed the efforts made by the institution to work with participatory approaches to agricultural technology development, and the ways in which these efforts had led them to interact with other institutions which, in the case of NGOs, included other NGOs and universities as well as public sector research and extension services. The bulk of these case studies were written, or directly supervised, by senior members of staff of these institutions.

At the same time, the case study papers were complemented with interviews, literature reviews, and country overview studies. Together these aimed to paint a more general picture of the changing social, political and economic contexts of NGO–government interaction in each country over the past two decades.

Throughout this process of documentation, the co-ordinators of the study themselves conducted much of the overview research, commissioned case study papers, and then worked closely with the individual authors of those studies. In addition, they helped to organize workshops at which the case study authors, and representatives from donor, government and NGO communities debated the themes raised by the different studies. A regional workshop for the Asia research was held in India – see Farrington and Lewis (1993), for Latin America in Bolivia - see Bebbington and Thiele (1993) and for West Africa in Senegal – see IIED (1992). In addition, country workshops were held in Kenya (Wellard *et al.* 1991), Zimbabwe (Wellard and Mema 1991), Bangladesh (Kashem *et al.* 1991) and the Philippines (Gonsalves and Miclat-Teves 1991).

Some definitions

While dealing with wider questions of institutional relationships, rural development and socio-political change, we have grounded our discussions in a particular theme: agricultural change and the development of agricultural technology. Within this analysis, our definitions of *agriculture* and of *technology* have been broad:

- *Agriculture* is defined to include crops, trees, animals and fish produced on-farm, and the interaction between the farm and the wider physical environment including, for instance, the collection of fodder or green manure from common land, and issues relating to the management of watersheds within which farms are located. The supply of inputs, and marketing and processing of agricultural products, are also included in this definition of agriculture, as are those backyard activities based on renewable natural resources which contribute to the livelihoods of the landless (e.g. livestock, but also, for instance, silk production);
- *Technology* is defined to include:
  - Hardware (e.g. seeds, vaccines, machinery);
  - Management practices and techniques (e.g. soil and water conservation practices, rotations, crop mixes, agroforestry);
  - Increments in knowledge, whether traditional, modern or some combination of the two that strengthen local capacity for experimentation, communication and resource management.

Several benefits were anticipated from our focus on the agricultural activities of these different institutions:

- It lends itself to empirical analysis and so goes beyond conceptual or rhetorical discussion of NGOs' supposed strengths and weaknesses;
- The types of NGOs in which we are primarily interested are either indigenous to developing countries, or are branches of northern-based NGOs. Within these categories, our interest is mainly in non-membership organizations which implement activities in their own right, or provide support to grassroots organizations such as community-based groups and co-operatives;
- Many of the rural poor rely on agriculture for large parts of their livelihood;

In seven of the eighteen countries discussed here, more than half of the labour force will still be engaged primarily in agriculture by 2025. In nine countries, the absolute size of the labour force engaged in agriculture will still be growing in 2025, implying increasing pressure on renewable natural resources. How to promote technical change in agriculture, and the institutional arrangements which support it, is therefore likely to remain a major area of policy formulation;

- Government efforts to develop new agricultural technologies and management practices for the rural poor have been characterized by only rare successes. Much of NGOs' work offers stronger prospects of success, but a key question addressed in this book is how these prospects might be implemented on a wider scale.

How to use this book

This book attempts to synthesize a wide range of empirical material from three continents. Inevitably, we had to be both brief and selective in choosing examples to illustrate our arguments. Readers wishing to know more about the case studies presented in the text boxes – and the contexts from which they derive – should refer to the companion volumes, one of which was produced for each Region.[1] The particularly busy reader may wish to turn to Chapter 6 for a summary of the main arguments and for our prognosis of likely future tendencies in NGOs' work and their relationships with governments and donors.

To state that the book is empirically-based is not to argue that it is solely empirical: we have been involved in teaching rural development to geographers, anthropologists, economists and sociologists, and on the basis of that experience have tried to produce a book that would fill a gap in the rural development literature, little of which deals analytically with the topic of NGOs. We have therefore subjected the contribution of NGOs to critical appraisal, locating the discussion in a broader empirical and theoretical literature on rural development, and relating it to the decisions that rural development managers have to make. At the same time, we have all worked closely with practitioners of agricultural and rural development, and have aimed to present the material in such a way that will give those people an overview of the diverse activities of NGOs, and the different ways in which they are, and might be involved in agricultural development activities.

We have aimed to draw out, as far as is possible, general patterns in the ways in which NGOs approach the development of agricultural technology, work with the rural poor, and interact with the state. This topic has taken us a long way from the initial focus on participatory technology development. The book also addresses the role of NGOs in poverty alleviation, the implications of structural adjustment programmes for the relationship between NGOs and the state, the contribution of NGOs to the process of rural democratization, and the ways in which NGOs do, do not, and might engage in policy dialogue.

We have defined the parameters of our task broadly. Consequently, we have generalized across a wide diversity of material, and have tried to bring several bodies of literature between the same covers – academic and applied literature from the social and agricultural sciences. As a result, there will be themes that certain readers feel receive insufficient attention. We make no apologies for that: there is much more research waiting to be done on NGOs, and our hope and intention is that this book will point to some of the questions that such research might pursue.

John Farrington
Anthony J. Bebbington
London, February 1993

NOTE

1 For Africa, see Wellard and Copestake (1993); for Asia, Farrington and Lewis (1993) and for Latin America, Bebbington and Thiele (1993).

ACKNOWLEDGEMENTS

By any standards, the production of this book has been an international endeavour. It is based on three years of research in eighteen countries. That research has brought us into contact with a large number of people, some of whom have been our collaborators in the preparation of case study material, others who have given us their time, thoughts and advice, and others who have simply been very good colleagues and friends. We cannot name them all here, though many are named in the separate regional volumes. None the less, we express once again our sincere thanks to them for their time, tolerance and insights.

This book is a companion to three regional volumes. We are particularly grateful to Kate Wellard, David Lewis, James Copestake, Graham Thiele, Martin Prager, Aurea Miclat-Teves, S. Satish, and Hernando Riveros for their support in producing the regional volumes.

The research on which the book is based was generously supported by the Overseas Development Administration, the International Development Research Centre, the Inter-American Foundation, the Ford Foundation, Winrock International and PRIP/PACT. In particular, we are glad to acknowledge the core support of a grant from the Natural Resources and Environment Department of ODA, which gave us the foundation on which to build this research programme, and a grant from the Ford Foundation which facilitated the writing up of all four volumes.

The text was written and produced at ODI; our thanks to colleagues at ODI and to its Director, John Howell, for support as we wrote the book. We would also like to recognize the contribution of the members of an informal group: David Brown, Alan Fowler, Mick Howes, Mark Robinson and Parmesh Shah who shared and commented on our ideas. Alison Saxby typed and organized more drafts than she would care to remember, and did so with exceptional patience, speed and accuracy. John Nelson and Simon Batterbury generously read and commented on the entire manuscript.

We are indebted to the large number of organizations – both NGO and governmental – which allowed their experiences to be documented, and, in many cases, allowed staff to take leave of absence in order to work with us.

ACKNOWLEDGEMENTS

Details of these organizations are given in the regional volumes, but the efforts of several in helping to coordinate aspects of the study merit particular acknowledgement. In Asia, the Administrative Staff College of India hosted the Asia Regional Workshop and provided the services of Dr S. Satish throughout the study; the National Academy of Agricultural Research Management, and its former Director Dr K. V. Raman, jointly coordinated the Workshop; the International Institute for Rural Reconstruction organized the Philippines Workshop; and Winrock International and Dr M. Hassanullah organized the Bangladesh Workshop. In Africa we collaborated with the International Institute for Environment and Development in organizing the West African Workshop, and Kengo and Voice/Nango organized the Kenya and Zimbabwe Workshops respectively.

In Latin America, the Centre for Tropical Agricultural Research (CIAT) together with the British Tropical Agricultural Mission (BTAM) in Santa Cruz, Bolivia organized a workshop jointly with ODI. BTAM also allowed Graham Thiele, Penny Davies and Jonathan Wadsworth to allocate time to the study; the Latin American Centre for Rural Technology and Education similarly assisted in the research and the workshop and allowed Martin Prager to dedicate time to the research.

While deeply indebted to these organizations and individuals, the authors alone remain responsible for the views expressed in this book.

GLOSSARY

ACDEP	Association of Church Development Projects (Ghana)
AGRARIA	AGRARIA: Food and Campesino Development (Chile)
AI	Artificial Insemination
AKRSP(I)	Aga Khan Rural Support Project (India)
ARAF	Fatick Region Farmers' Association (Senegal) (Association Regionale des Agriculteurs de Fatick)
ATA	Appropriate Technology Association (Thailand)
ATD	Agricultural Technology Development
ATS	Agricultural Technology System
AWS	Action for World Solidarity (India)
BAIF	Bharatiya Agro-Industries Foundation (India)
BRAC	Bangladesh Rural Advancement Committee
CAAP	Andean Centre for Popular Action (Ecuador) (Centro Andino de Acción Popular)
CAPART	Council for the Advancement of People's Action and Rural Technology (India)
CARE	American Co-operative for Remittances to the Exterior
CARE-Lift	CARE–Local Initiatives for Farmers Training (Bangladesh)
CATT	Technology Adjustment and Transfer Centres (Chile) (Centro de Ajuste y de Transferencia de Tecnología)
CEDLA	Centre for Labour and Agrarian Studies (Bolivia) (Centro de Documentación Laboral y Agraria)
CELATER	Latin American Centre for Rural Technology and Education (Colombia) (Centro Latinamericano de Tecnología y Educación Rural)
CEPLAES	Centre for Planning and Social Studies (Ecuador) (Centro de Planificación y Estudios Sociales)
CESA-Bol	Centre for Agricultural Services (Bolivia) (Centro de Servicios Agrícolas)
CESA-Ec	Centre for Agricultural Services (Ecuador) (Central Ecuatoriana de Servicios Agrícolas)
CGIAR	Consultative Group on International Agricultural Research

CIAT	Centre for Tropical Agricultural Research (Bolivia) (Centro de Investigación Agrícola Tropical)
CIED	Centre for Research, Education and Development (Peru) (Centro de Investigaciones, Educación y Desarrollo)
CIPCA	Centre for Campesino Research and Development (Bolivia) (Centro de Investigación y Promoción del Campesinado)
CLADES	Latin American Consortium for Agroecology and Development (Chile) (Consorcio Latinamericano de Agroecología y Desarrollo)
CONAF	Corporación Nacional de Forestería
CRAR	Resource Centre for Regenerative Agriculture
CRS	Catholic Relief Service
CUSO	Canadian Universities Service Overseas
DENR	Department of Environment and Natural Resources (Philippines)
DTT	Department of Technology Transfer (located in CIAT) (Departamento de Transferencia de Tecnología)
El CEIBO	The El Ceibo Regional Agricultural and Agroindustrial Co-operative Co-ordinating Committee (Bolivia)
ENDA	Environment and Development Activities
ERDB	Ecosystems Research and Development Bureau (Philippines)
EXTIE	International Economic Relations Division (of the World Bank)
FESD	Forestry Extension Services Division (Kenya)
FEVORD-K	Federation of Voluntary Organizations for Rural Development in Karnataka (India)
FITT	Farmer Innovation and Technology Testing Programme (The Gambia)
FIVDB	Friends in Village Development, Bangladesh
FSRI	Farming Systems Research Institute (Thailand)
FUNDAEC	Foundation for the Application and Teaching of Science (Colombia) (Fundación para la Applicación y Enseñanza de la Ciencias)
GDP	Gross Domestic Product
GIA	Group for Agrarian Research (Chile) (Grupo de Investigaciónes Agrarias)
GO	Government organization
GONGO	Government-organized NGO
GRO	Grassroots organization
GSM	Good Seed Mission (The Gambia)
GSO	Grassroots support organization
GVAM	Gwembe Valley Agricultural Mission (Zambia)

IAF	Inter-American Foundation
IARC	International Agricultural Research Centre
IBTA	Bolivian Institute for Agricultural Technology (Instituto Boliviano de Tecnología Agropecuaria)
ICAR	Indian Council for Agricultural Research
ICRAF	International Centre for Research in Agroforestry
IDEAS	IDEAS Centre (Peru)
IFAD	International Fund for Agricultural Development
IIRR	International Institute for Rural Reconstruction (Philippines)
INDAP	Institute for Agricultural Development (Chile) (Instituto de Desarrollo Agropecuario)
INIA	Institute for Agricultural Research (Chile) (Instituto de Investigaciones Agropecuarias)
INIAP-PIP	Institute for Agricultural Research – Production Research Programme (Ecuador) (Instituto Nacional de Investigaciones Agropecuarias – Programa de Investigación en la Producción)
IPM	Integrated pest management
ISNAR	International Service for National Agricultural Research
ISRA	Senegalese Institute for Agricultural Research (Institut Sénégalais de Recherche Agricole)
ITK	Indigenous Technical Knowledge
IU	Intermediate users (of technology)
IUCN	International Union for the Conservation of Nature
KEFRI	Kenya Forestry Research Institute
KENGO	Kenya Energy and Environment Organisation
KVK	Krishi Vignan Kendra (Farm Science Centre) (India)
LP3ES	Institute for Social and Economic Research, Education, and Information (Indonesia)
MBRLC	Mindanao Baptist Rural Life Center (Philippines)
MCC	Mennonite Central Committee (Bangladesh)
MFI	Maguugmad Foundation Inc (Philippines)
MSO	Membership support organization
MYRADA	Mysore Relief and Development Agency (India)
NAF	Nepal Agroforestry Foundation
NARS	National Agricultural Research Service
NGO	Non-governmental organization
ORAP	Organization of Rural Associations for Progress (Zimbabwe)
PHS	Protected Horticultural Systems
PIDOW	Participative Integrated Development of Watersheds (India)
PO	People's organization

PRA	Participatory Rural Appraisal
PRADAN	Professional Assistance for Development Action (India)
PROCADE	Programme for Alternative Campesino Development (Bolivia) (Programa Campesino Alternativo de Desarrollo)
PRRM	Philippine Rural Reconstruction Movement
PVO	Private voluntary organization
RDRS	Rangpur Dinajpur Rural Service (Bangladesh)
REFORM	Resource and Ecology Foundation for the Regeneration of Mindanao (Philippines)
RKM	Ramakrishna Mission (India)
SEPAS	Diocesan Secretariat for Pastoral and Social Work (Colombia) (Secretariado Diocesano Pastoral Social)
SMAP	Southern Mindanao Agricultural Program (Philippines)
SMS	Subject matter specialist
SSNCC	Social Services National Co-ordination Council (Nepal)
SWDC	State Watershed Development Cell (India)
TAAP	Tamale Archdiocese Agricultural Programme (Ghana)
UMN	United Mission to Nepal
UPASI	United Planters' Association of Southern India
WARDA	West African Rice Development Association
WDP	Watershed Development Programme (India)

1

INTRODUCTION:
MANY ROADS LEAD TO NGOs

After several decades in the wings of development practice and debate, non-governmental organizations – NGOs – have quickly moved to centre stage. The explosion of interest in them has come from different quarters: from academic researchers, development activists, multi- and bilateral donor agencies, and not least from society itself. In academia this interest is reflected in research programmes and a growing body of published work focusing on NGOs. Society's interest is reflected in the rising contributions to NGOs, and the growing frequency with which representatives from them are interviewed in the mass media. Some donor agencies, such as the World Bank, now have departments specially responsible for NGOs, and departmental statements that emphasize work with NGOs. Certainly they and northern governments are channelling ever greater amounts of money through the non-governmental sector.[1]

In this introductory chapter, our purpose is to define the term 'NGO' and to chart some of the experiences and changes in development thinking that have led to this enthusiasm for NGOs. We will pick out certain lines of reasoning about the state, about civil society, and about technology, that have led analysts and donors alike to the non-governmental sector. Our specific analysis in the remainder of the book deals with NGOs and agricultural development activities, and part of this chapter deals with issues that have been specific to the agricultural sector. However, much of the interest in NGOs stems from debates on development theory and policy, and so much of this chapter is given over to this wider arena.

As we will see, the reasons for this surge of interest are diverse, and not always mutually compatible. For some, NGOs are in the vanguard of an alternative mode of development that is fundamentally different from today's neo-liberal orthodoxy; other lines of reasoning see NGOs playing roles within the existing neo-liberal framework. Many roads lead to NGOs.

It would be convenient if we were able to apportion these different perspectives on NGOs to specific institutions. Unfortunately, such analytical tidiness is impossible. Similarly, there are no easily definable, mutually exclusive neo-liberal, or post-marxist positions on what NGOs should contribute to development. In part this is indicative of the diverse views that

1

tend to coexist within any one institution or school of thought. But it is also the effect of a shared uncertainty about how to define, and then implement, successful development. Thus, in an institution, such as the World Bank, that some might define as a bastion of neo-liberalism, we can also find an increasing willingness to argue that development should accord as much importance to human rights and democratic process as to economic growth (EXTIE 1990). Among radicals, the lines of the development debate are now far less clearly drawn than they were when 'modernization and modes of production' were counterposed (Taylor 1979; cf. Booth 1992). There is now greater willingness to recognize the potentially virtuous contributions of the private sector to development, the limitations of the state as a vehicle of progressive social change, and the need for serious consideration of economic efficiency in the delivery of development services.

At the heart of this blurring of lines is a need felt by many to rethink existing concepts of the state and the market, and what they can and should contribute to development. Experiences with both are mixed (Colclough and Manor 1991; Uphoff 1993). The excesses of state inefficiency, repression and corruption require a rethinking among those who previously assumed that socialism, or at least social development, would be achieved through public sector actions. On the other hand, nor have profit minded actors in the market shown much willingness to eradicate poverty, empower the poor, or even to invest productively in the wake of neo-liberal economic programmes.

This rethinking has stimulated reformed radicals and neo-liberals alike to look for a 'third sector' (Korten 1987) to alleviate poverty, strengthen civil society and promote efficient and participatory grassroots development in ways beyond the capability – or willingness – of the state and the market (Uphoff 1993).[2] This 'third sector' has been found in the complex of voluntary, self-help and non-governmental organizations that have long been present in society. Some analysts come to these NGOs more interested in poverty alleviation and efficiency, others more concerned with empowerment and a stronger civil society, and others with an interest in 'green' development. They therefore place particular stress on different roles this third sector can play in development. But, as we shall see, the different perspectives also have much in common.

The uncertainty in development thinking is played out in a parallel uncertainty and multiplicity of views about what NGOs should contribute to development, as we shall suggest below. A specific aspect of this uncertainty – and one which was bound to arise from a fundamental reconsideration of the roles of state and market – concerns the relationship of NGOs to the state. We will return to this at the end of the chapter. First, however, after making a few comments on the definition of NGOs, we chart several areas of debate that have identified special roles for NGOs in (i) reforming the role of the state in development, (ii) supporting the self-managed development actions of grassroots groups, and (iii) making agricultural technologies more widely available, and more environmentally sound.

2

WHAT ARE NGOs?

Many roads may be signposted 'NGOs,' but there is considerable confusion in both the literature and among policy makers as to what we mean by the label NGO (Munck 1992). Authors such as Clark (1991) tend to use the label NGO all inclusively. Clark (1991: 34–5) distinguishes six categories of organization (relief and welfare agencies, technical innovation organizations, public service contractors, popular development organizations, grassroots development organizations and advocacy groups and networks), but calls them all NGOs. As he himself says, such all inclusiveness can make the term almost meaningless. Part of the problem is that the classification does not fully differentiate between the function, ownership and scale of operation as part of a sub-categorization of these organizations. As a result, everything from a neighbourhood organization concerned with better lighting through to an organization operating globally, such as Oxfam, are equally labelled 'NGO'.[3]

A first step in classifying these groups is to distinguish according to their origins (Figure 1.1). Are they northern NGOs that have activities in the South, or the southern-based branches or affiliates of these that operate with a high degree of autonomy, or indigenous South-based organizations? Within these last two categories, there are grassroots organizations (communities, co-operatives, neighbourhood committees, etc.), organizations that give support to the grassroots, and those that engage in networking and lobbying activities (these functions are not necessarily mutually exclusive). Among these, some are North-owned (e.g. field-based Oxfam programmes), others are indigenous to the South. Most of our material deals with indigenous groups, particularly in Latin America. North-based organizations are proportionately more significant (and national NGOs correspondingly less so) in Africa, and so the African material presents more evidence on North-based groups operating in Africa.[4]

Among the organizations in the South, it is important to distinguish between them according to the nature of their relationship to the rural poor, and their staff composition. On the one hand are those organizations that are staffed and elected by the people they are meant to serve and represent (such as farmer organizations). Following Carroll (1992)[5] and Fowler (1990), we call these membership organizations. Non-membership organizations are, by contrast, staffed by people who are socially, professionally and at times ethnically different from their clients. Similarly, the two types of organization have different relationships with the poor who are members of the former, and clients of the latter.

Membership and non-membership organizations can each engage in both advocacy activities and development actions in which they give services to the grassroots. Many combine the two sorts of action, though generally specialize in one or the other. Again, our emphasis has been on the service organizations,

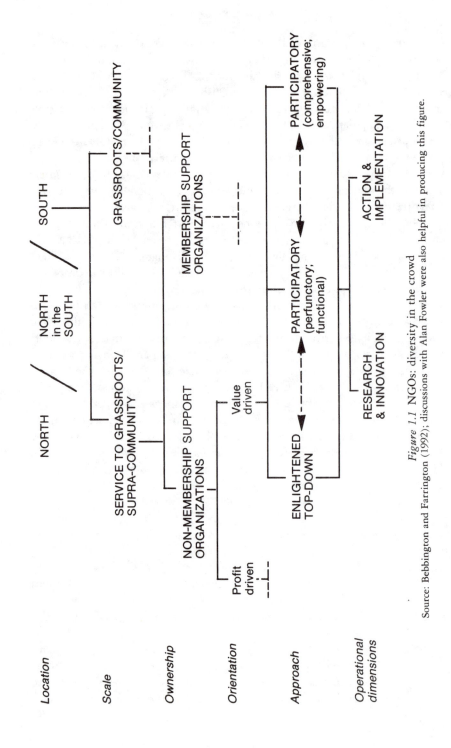

Figure 1.1 NGOs: diversity in the crowd

Source: Bebbington and Farrington (1992); discussions with Alan Fowler were also helpful in producing this figure.

although some such as CLADES and KENGO are more engaged in advocacy and networking activities.

Talking about the organizations that give services to, or lobby on behalf of their clients, Carroll (1992) combines questions of function and ownership, and distinguishes between 'membership support organizations' (MSOs) which are staffed and elected by their clients, and 'grassroots support organizations' (GSOs) which are professionally based and in some sense independent from grassroots control. This distinction is important, for the two sorts of organizations have differences in a number of dimensions, including:

- Claims regarding how far they represent grassroots concerns;
- Their management styles and stability (elected staff are more likely to change);
- Their professional competence;
- The directions in which they are accountable (especially the difference in how far they are accountable to donors and the rural poor);
- The social networks to which they have easy access and from which they are able to draw down resources and contacts.

There is a strong a priori case, often put emotively, to be made for working with membership organizations if democratic forms of grassroots development are our goal. We too would endorse that idea as a long-term goal, and have written about it elsewhere (Bebbington *et al*. 1992; 1993). Similarly, many GSOs state that their primary goal is to strengthen such organizations and ultimately pass control of projects to them. However, for reasons we will elaborate below, there are also grounds for guarding against optimism about membership organizations (cf. Carroll 1992). It is for this reason that in this book we have focused primarily on GSOs, although as we use the term 'NGO' in the text, our main referents are both GSOs and MSOs.

RETHINKING THE STATE

Much of the interest in NGOs has been generated by a disappointment in the past performance of the state. This poor performance has had economic and political dimensions. There have been economic concerns about the inefficiencies created by the state's intervention in the economy, including its implementation of development programmes. Equally, there have been political concerns that many states have not been accountable to society, and indeed have been more interested in controlling and moulding society to suit their own interests, than in responding to the needs of that society.

Problems of efficiency

Traditionally, proposals for development programmes have assumed that the state and its many agencies were the vehicles through which projects and policies would be implemented – even if the understanding of how the state operated was often both weak and naive (Long 1988). The dominance of such state-centred thinking and action originated in some cases from Keynesian and import-substitution models of development in which the state was given the role as main protagonist in seeking to expand domestic markets and domestic capacity for industrial and agricultural production, and in breaking the dependency on export markets. In other cases, particularly in Africa, the state became the protagonist in the post-colonial project of nation building and self-determined development.

In the agricultural sector, the effect of this state-centred strategy was to litter the institutional landscape with parastatal marketing companies, agrarian banks, land reform agencies, public institutions for agricultural research and extension, irrigation agencies, rural development programmes and many more. If field researchers wanted to know about development programmes they almost instinctively went to the state agencies.

This growth of the state and the proliferation of its institutions (Martínez Noguiera 1990) brought with it a number of inefficiencies that caused growing concern to policy makers, and particularly to donor agencies – even though it had been those same agencies who in earlier years supported these institutional developments. Overall the growth of the state and struggles to influence its decisions led to inefficient allocation of resources at a national level, and particularly between rural and urban sectors, and private and public sectors (Krueger 1976; Lehmann 1990). The cost of sustaining the state structure, aside from implementation costs, diverted resources from other potentially more productive uses. Furthermore, many argued, decision makers in those institutions made policy choices that were urban-biased, motivated by institutional, political and even kinship interests, and hence had negative impacts on the rural sector (Bates 1981; Corbridge 1982; Grindle 1986; Hyden 1983; Lipton 1977).

In addition, many argued, the intervention of the state – in markets, pricing policies, and programme implementation – brought further inefficiencies (Krueger et al. 1991). Parastatals that were not subject to market forces had excessive operating costs, and their payments to producers were typically erratic in timing and quantity. This in turn generated further costs, as producers dedicated resources to evading the system (e.g. by smuggling) or simply withdrew from the market (Hyden 1983). In rural development projects, the tendency for state institutions to centralize decision making led to growing classes of urban-based functionaries, hierarchical decision-making and so reduced flexibility and responsiveness. This easily led to inappropriate and simply slow programme implementation at a local level (Barsky 1990). The biting crisis of public sector finances in the 1970s and 1980s aggravated these problems as resources were spread ever more thinly leaving public institutions without the materials and funds necessary simply to sustain operations. Public sector agricultural research institutes began to cut funding

6

for field work, and supplies of materials for work-on-station remained unreplenished.

These difficulties were compounded by personnel problems. Institutions were often used for political patronage, to deliver resources to politically favoured areas, and to give jobs to party members and other clients (Hyden 1983). The problems of low motivation and commitment that inevitably arose were compounded by falling real wages in the public sector which led many to leave their jobs, take long periods of paid leave, or simply use the jobs as a basis on which to do other work as consultants (or taxi drivers).

In the 1980s, donors not merely withdrew from the earlier models of state-centred development; they made them the object of attack by an armoury of policy reforms. As part of the general packages of structural adjustment, the IMF and the World Bank have set the trend by demanding public sector reforms centred on reduced levels of expenditure, public sector restructuring, and state withdrawal from market and indeed project interventions (Moseley et al. 1991). Many bilateral donors linked their aid to these reforms – governments had to bite the multilateral bullet before receiving any bilateral bandages.

Before we demonize the donors too much, however, it ought be said that public sector retrenchment often merely formalized a de facto situation in which the state had already collapsed and was doing little or nothing of significance for the middle and lower income groups. Either way, these policy changes stimulated the search for new mechanisms of service delivery.

In certain instances, the market becomes the new mechanism – as in the privatization of parastatals. However, in agricultural development for resource-poor farmers this has been a less viable option. It is even less so in programmes specifically concerned with the alleviation of poverty and of the social effects of structural adjustment. Indeed it has been in these areas that donor agencies have made one of the most concerted initiatives to channel development (as opposed to emergency) funds through non-governmental organizations. These were formative experiences for the donors. The relative success of involving NGOs in the design and implementation of small projects financed by the Social Emergency Fund in Bolivia, for instance, was significant in the World Bank's subsequent thinking about involving NGOs in such funds and other projects (Williams 1990; World Bank 1990c).

In the agricultural sector, these public sector reforms have not always stimulated attention to alternative institutional arrangements, and the public sector has simply been left to get weaker and weaker by itself, without any attempt to involve other organizations. In some cases, however, policy changes have had a direct impact on research and extension systems and on the ways in which strategies and services for rural development might be organized and institutionalized. The exact form of that impact has often been determined by the answers to a series of policy questions that these changes are forcing:

7

- What functions should continue being those of a scaled down public sector?
- What activities and responsibilities should be passed over to the private sector?
- What sort of agent in that private sector should be expected to assume those responsibilities – commercial interest, community organization or non-governmental organization?
- What incentive, if any, will those private agents need to encourage them to assume these new roles?
- How might a reduced public sector best interact with these private agents?

The answers to the questions have varied between countries, generally in relation to the resources at their disposal. Thus, in some instances donors and governments have concluded from their review of the situation prior to these policy changes that the state had ceased to perform any significant role in one or other development activity. At the same time they have observed the extent to which the non-governmental sector has assumed these roles and become the only effective vehicle of agricultural development. On the basis of these observations some have simply decided to formalize the situation and say, as in the case of Bolivia for instance, that the public sector institution previously responsible for agricultural research and extension, will in the future dedicate itself only to research, and explicitly leave extension to NGOs and farmers' organizations. The link between research and extension will thus become a link between the state and NGOs (Bebbington and Thiele 1993). In other cases, as in Chile, where the state is relatively well resourced, it has retained responsibility for setting and financing extension programmes, but has contracted their implementation out to the private sector, allowing commercial, non-governmental and farmer organizations the right to bid for these contracts.

The reasons for moving to involve NGOs in these ways are many, but it is clear that much of the motivation is pragmatic, and the role perceived for them is instrumental. Donors want them to do the jobs the state previously did, or was supposed to have done. The attraction is that the NGO does not represent a recurrent cost for government, and so to channel programmes through the non-governmental sector will not build in an increase in public expenditure – unlike the case where state institutions and their staff continue to absorb public resources beyond the period of the project that created them. Furthermore, channelling a project through an NGO may allow the donor to take advantage of the NGO's institutional structure and so gain an element of subsidy from the NGO. Finally, it is assumed (though not proven) that NGOs are more efficient in their use of funds, so implementing programmes through them will give better value for money.[6]

Problems of accountability: civil society and the state

Many of these inefficiencies and biases in resource allocations by the state are not merely the effects of bad decisions but, more fundamentally, consequences of the weakness or absence of mechanisms through which society can hold the state accountable for its actions, and demand from it policies and programmes which meet societal needs. A recent survey of the literature on governance and democracy in Sub-Saharan Africa concluded that the general experience of the African state has been quite different from that of the liberal western *model* of the state as a neutral arena in which interest groups seek to influence policy and resource allocation. Rather, the African state has typically been controlled by small elites. Healey and Robinson conclude:

> The apparently irrational policies of the 1970s seem to be partly explained by the following features. Firstly, policies have been made in small, exclusive political circles . . . with little consultation of societal interests . . . Second, the most powerful influence seems to have been the search for political legitimacy . . . through a patronage and clientelistic process rather than though open politically competitive electoral systems. . . . A third but weaker influence on policy was the regime's ideology.
>
> (Healey and Robinson 1992: 90)

In relation to Latin America Lehmann similarly concludes

> Governments in Latin America . . . have been able . . . to make radical departures in economic policy which are unthinkable in either capitalist or socialist advanced industrial societies, where established habits and interests are more deeply entrenched.
>
> (Lehmann 1990: 204)

Such situations of extreme state neglect of societal concerns are often due in part to the political and organizational weakness of civil society, and in particular of the rural poor. The significance of this factor is demonstrated by the determination with which authoritarian and rightist regimes in Chile and Bolivia sought to destroy peasant organizations, which had in earlier periods been strong enough to gain policy concessions as significant as land reform (Thiesenhusen 1989; Lehmann 1990; Carroll 1964). In Africa the peasantry has been consistently far weaker as a political force, and this has been explained by several commentators as the cause of its inability to affect general and specifically agricultural, state policies (Hyden 1983; Bates 1981; Healey and Robinson 1992). Drawing on the last decade's experience of a North American NGO, Gubbels (1992) has recently asserted that there is little likelihood that a farmer-first research and extension strategy is going to materialize in West Africa as a result of peasant pressure – the peasantry is simply too weak. However, while these general observations are helpful for

9

our understanding, it is important to recognize that the institutional solidity of civil society does vary among different African countries, and indeed while the peasantry may be a weak pressure group, there are countries in which larger farmers have grown into strong interest groups. The delay of land reform in Zimbabwe (Wellard and Copestake 1993) is a classic example of how large farmers have influenced the state – to the detriment, incidentally, of their smaller colleagues.

The weak accountability of government can also be due to state actions that are deliberately intended to prevent such strengthening in civil society and to remove other mechanisms through which public institutions may be held accountable – such as the free Press and the system of multiparty electoral democracy. Such situations existed until the 1970s and 80s in many countries in Africa, Asia, Latin America, and indeed in Europe. In some countries they still exist. In the worse cases in Latin America they have been typified by violent repression of the political left and all assumed to support it – peasant organizations, unions, intellectuals in university departments, and indeed NGOs. In Africa this ideological axis of repression has been further overlain by ethnic questions. In less extreme circumstances such repression is more mundane and less violent, and manifested in a series of legal and institutional means of undermining civic association.

In recent years donors have finally begun to worry about these questions. In part this is because they recognize that the models of development they wish to foster will not be consolidated unless effective demand for them can be articulated by the relevant sections of the society. In part it is because the values of human rights and freedom have been recognized as part of development – as something that should not be sacrificed in the name of growth. And, in a cynical vein, donors now perceive democracy as something to be promoted now that the cold war has ended and there is less political need to keep certain Third World tyrants in power.

Donors' discovery of democratization and governance as goals of development has led to the search for means to promote them. One means has been political conditionality, as witnessed in decisions of the G-7 to block aid to Malawi and Kenya until they satisfied certain standards of freedom of Press, speech and assembly (ODI 1992). At the same time donors have sought to use development projects to strengthen civic associations and create new mechanisms through which state institutions can be held to account.

It is in this vein that the governance debates have led donors to NGOs. Many have suggested that if NGOs are strengthened, by channelling funds through them and by increasing their control over development programmes, so too will be civil society as a whole. Similarly, projects which aim to create decision making processes in which NGOs have a presence allow new means through which civil society can exercise influence over public decisions, and of opening that public process to more scrutiny.

In many respects, the increased concern of multilateral agencies to incorporate NGOs' views into projects which have major implications for the environment and for human displacement, is a similar instance of their attempts to create mechanisms through which the NGOs can make demands on the state (Williams 1990). In all these ways, NGOs have been deemed representatives of the rural poor, though as we will see later, this is itself a problematic assumption.

At the same time multilateral and bilateral agencies have identified NGOs as institutions that can strengthen local organizations and promote popular participation in the development process (World Bank 1991). In this regard, strengthening and involving NGOs can have, supposedly, a doubly positive impact in the consolidation of civil society. They are part of civil society, and at the same time their work can strengthen other institutions of that society – particularly organizations of the rural poor.

It is not simply the donors who have discovered civil society and democratization. These themes have likewise become more prominent in academic debates on development. Several important texts have recently addressed development and democracy together (Lehmann 1990; Friedmann 1992; Healey and Robinson 1992; Fox 1990a). Once again there must be many reasons for this, but the rethinking of Marxian perspectives has been an important factor, perhaps especially in Latin American discussions.

Theoretical debates have responded to the apparent failures of both class analysis and class action to predict and achieve change, with the steady decline of class-based organizations and the parallel rise of 'new social movements' grounded in identities related to gender, place, and ethnicity. The apparent strength, ubiquity and not-so-newness of these movements has turned them into a focus of a new literature (Slater 1985b; Harvey 1991; Redclift 1988). This literature has emphasized that the movements are characterized by the concern to demand rights of citizenship from the state, to express social identities not catered to by class organizations, and to express and exercise a resistance to the many ways in which state and wider society have dominated popular sectors. These observations generate a vision of civil society as a far more complex mosaic of different interest groups than class and dependency analyses had painted. At the same time, they are less revolutionary than the analyses of the 1960s and 1970s, and rather than emphasize the takeover of the state, they stress the need to make the state more transparent, open to public scrutiny and more responsive to societal concerns. This in turn leads to the argument that these social movements must be strengthened if this is to be achieved.

NGOs enter these discussions about social movements and civil society in different ways. Some seem to suggest that they are movements in themselves, but most argue that they constitute an interest group in society that at the same time provides support to grassroots organizations.

Their role as 'grassroots support organizations' (a term attributable to Carroll 1992) has received particular attention as a force for the consolidation of ciivl society, not least in rural areas. The history of agrarian movements has suggested that, even in those areas where they have been at times strong (such as Latin America), they suffer a number of institutional weaknesses that range from limited management and administrative skills, to vulnerability to capture by leaders and elites, to ephemerality (Lehmann 1990; Bebbington *et al.* 1993). These limitations weaken the possibility that these movements will be able to sustain political pressure on the state – as the organization weakens, so too does its effectiveness as a pressure group. As long as rural organizations come and go, the state is more able to weather their pressure. The implication is that the effectiveness of civil society requires that these organizations themselves be strengthened with significant and targeted external support. This, several have argued, is the key contribution that NGOs can make. By working with grassroots groups in diverse capacities – such as training, or the joint selection, implementation and monitoring of projects and programmes – by stimulating contact among such groups, and by facilitating their creation where they do not already exist, the best NGOs can play an important role in ensuring the survival, maturation and indeed internal democracy of those organizations (Fox 1990; Carroll 1992).

CIVIL SOCIETY AND SELF-MANAGED DEVELOPMENT

The activities of the institutions of civil society has attracted the interest of analysts and donors not only for their potential to make the state more accountable, but also for their role as development institutions. For as well as being political agents, many of these new agrarian movements have assumed a self-help and developmental role (Bebbington *et al.* 1993; Harvey 1991). They design their own projects and negotiate the funds they require from international donors.

This is not a new phenomenon – some funding agencies, such as Oxfam and the other non-governmental donors in the North, have long supported these sorts of grassroots initiative. What is new is the attention that they have come to attract from those analysts who write about alternative development, and grassroots and farmer-first approaches (Annis and Hakim 1988; Chambers *et al.* 1989), and from other bi- and multilateral donor agencies who in the past were little concerned with such local organizations (World Bank 1992). Both have begun to look more closely at these initiatives, and have concentrated on their socio-political significance as well as their practical effectiveness.

For authors such as Friedmann (1992) these self-help actions represent an attempt on the part of the grassroots to assert greater control over the environments in which they live. Friedmann contends that local self-development actions contribute to a process of collective self-empowerment in two senses. Psychologically and organizationally they begin to build grassroots capacity to take hold of local development problems – an issue to which Carroll (1992) also apportions paramount importance. At the same time, they have a more obvious economically empowering role in reducing material poverty.

Friedmann's analysis of the role of NGOs in this process is interesting because it approaches them from a theoretical reflection on the nature of development. Friedmann's own conception of an 'alternative development' aims to mould a middle ground between economistic, Marxian and populist approaches to development. To the first two he adds a dose of humanism and civil rights. Development should not only be about growth, he argues, but about the construction of a society that gives people more power to fashion and enjoy the spaces in which they live. It should broaden democracy beyond the (not so) simple right to vote in elections, into a direct democracy that would, in Norberto Bobbio's (1987) terms, increase the range of contexts outside politics where people can exercise the right to vote. It must also take on board issues of gender, democracy and sustainability. On the other hand, while he insists that the theory and practice of development must recognize and develop human agency, populist formulations that concentrate only on consciousness raising, local knowledge and grassroots action must be more politically realistic. They must find ways to link local action with strategies to remove structural obstacles to human development.

This is where his argument turns to grassroots organizations and those NGOs which give them support. These are, he asserts, the organizations that can empower people both economically and psychologically, developing their capacities and determination to claim rights. They also become vehicles, albeit imperfectly so, for helping to link local action back into national and structural change. For Friedmann then, as for David Lehmann (1990), local organizations, membership and non-membership, inspire the thinking behind a theoretical reformulation of development and become the vehicles through whose actions that possibility might be made real.

Contributions such as Friedmann's and Lehmann's show the academic literature beginning, belatedly, to respond to the reality of grassroots organizations' and NGOs' work in rural development. These analyses begin to link this NGO activity to debates on democracy, development and civil society. This academic response has been slower than that of those development agencies who earlier on realized the significance of this NGO activity.

While these academic concerns overlap with donor concern for governance and democracy, the perspectives on NGOs that have perhaps been more persuasive in the development agencies have been those emphasizing that these grassroots actions are more efficient in resource use and more effective than those of the state in reaching the poorest strata of society, and in promoting the participation of poor people in project design, implementation and monitoring. Work such as Clark's (1991), Uphoff's (1986), Carroll's (1992) and Annis and Hakim's (1988) have played, or continue to play, important roles in furthering these arguments.

As these assertions about NGO effectiveness and efficiency have passed into popular lore, and as bilateral and multilateral donors become once again preoccupied with poverty alleviation (World Bank 1990a), so these agencies have sought to channel more funds through NGOs. Furthermore, it is more openly recognized now that in certain circumstances, especially in Latin America, but also in some S. Asian and African countries, where the state was extremely weak, deliberately ignored the needs of the rural poor, or had unacceptable political agendas, it has been NGOs who have been doing the bulk of development work at the grassroots.

However, the enthusiasm for 'NGO' contributions to self-managed development has often operated with very loose conceptions of what NGOs are. Analysts such as Clark (1991) and Cernea (1988), both now of the World Bank, refer to grassroots groups, service organizations, and international non-governmental agencies all as NGOs. This has analytical shortcomings, for it fails to address the quite differing ways in which these organizations are part of civil society. For donor institutions it has a further strategic shortcoming, for once it is decided to channel funds towards 'NGOs', the agency then must decide to which type of NGO funds ought be channelled, and why.

In responding to these practical concerns, many donors have preferred to channel funds via intermediary agencies, often non-membership NGOs or special government agencies such as social funds. This preference reflects the logistical impossibility for donors of working with large numbers of grassroots groups.

The question of whether intermediaries should be used, however, ought also be assessed on their potential contribution to development. In this regard there is little agreement. For those who would rather channel funds 'direct to the grassroots' (Annis and Hakim 1988), the presence of intermediary NGOs can often be seen as an unnecessary drain on development resources. Such NGOs are generally urban-based, sustain offices and management infrastructure, and are different in social class and ethnicity from the rural poor they aim to serve. Funding them thus benefits people who are most definitely not the poorest of the poor. This criticism has gained more force, and credence, in recent years as these NGOs have proliferated in response to the growing interest among governments and donors to work with NGOs. Some of these NGOs are clearly more a new income generating strategy for middle class professionals feeling the pinch of structural adjustment than they are development institutions.[7]

On the basis of these observations, some profess a preference to go straight to the grassroots letting them build on their own knowledge, oriented to their own felt needs (Rivera-Cusicanqui 1990; Kleemeyer 1991; 1992).

However, this populist interpretation of the potential of grassroots groups to manage their own development is itself problematic. The poorest of the poor are often the least able to bear the costs implied by organizing, and so to work with existing organizations may by-pass them. Furthermore, groups that exist are not necessarily as skilled at management, or as internally democratic as some would wish to believe (cf. Slater 1985a). Carroll (1992) has drawn attention to a number of myths about this grassroots capacity to organize and manage development. Some of these myths, he mischievously suggests, stem from populist perceptions in donor agencies themselves. The belief in the innate capacity of the popular sectors to organize themselves without external support he calls a myth of 'spontaneous combustion' and 'immaculate conception'. He also warns that grassroots groups are vulnerable to takeover by small interest groups, leading to biases in their distribution of resources and also implying that when they present themselves to donors as 'community organizations', community-wide they are not. Indeed, a long literature since Landsberger and Hewitt's (1970) 'ten sources of weakness and cleavage in peasant movements' suggests that the question of the internal democracy and management capacities of such membership groups is vexed and complicated.

Carroll implies that we do the grassroots no service at all by overstating their capacities. Indeed, against those who insist that donor support should go direct to the grassroots, or at least to MSOs, Carroll argues that GSOs perform better than MSOs on all his criteria of evaluation. These results lead him to suggest that much more attention should be paid to building capacity in the popular sectors – i.e. developing their capacities for collective self-management and for informed political negotiation with external institutions. GSOs and MSOs must improve grassroots groups' ability to do this work. For Carroll, this is ultimately their most crucial contribution to grassroots development. In their survey of local organizations, Esman and Uphoff (1984) come, by implication, to the same conclusion – that base organizations cannot easily acquire the same skills as GSOs. These issues are considered more fully in our discussion of NGOs' interaction with their clients in Chapter 4.

Carroll therefore endorses the idea that intermediary NGOs should receive funds in preference to the grassroots, a conclusion that Lehmann (1990) also arrives at. However, Carroll's results are also a warning to over-enthusiasm about these sorts of NGO. For despite his conclusion that they have a special contribution to make to building local capacity, his findings show that, for all their talk about participation and capacity building, NGOs perform better at delivering services (inputs, seed, health, education, etc.) or implementing projects such as road and canal-building. Carroll's work in Latin America also comes to the same conclusion as Riddell and Robinson's (1993) in Africa and Asia – that NGOs do not reach the poorest of the poor. Chapter 4 provides further discussion of NGOs' interaction with the poor.

In this book we too have concentrated on intermediary NGOs, on similar grounds – that we feel they have a special contribution to make to strengthening self-managed development. Ultimately, *if* they build capacity they will hand more and more administrative tasks over to membership groups, and either disappear or identify new types of support role for themselves. First, however, they must build capacity.

15

HIGH EXTERNAL-INPUT AGRICULTURAL TECHNOLOGY: SPREADING IT AND QUESTIONING IT

These debates about state and civil society, about efficiency and poverty alleviation, have led to a general increase in interest in NGOs. In this book we are dealing with NGOs in agricultural development, and in particular their role in developing agricultural technologies for the rural poor. And indeed, there are also some specifically agricultural debates that have inspired increased interest in the potentially special roles of NGOs.

These debates stem in large measure from the on-going re-evaluation of the high external-input approach to agricultural development that, according to popular conception, characterized Green Revolution strategies.[8] Of particular importance have been the debates as to the relevance of this technology to small farmer needs, and its environmental sustainability.[9]

Adapting agricultural modernization strategies

The debate as to whether the rural poor were able to benefit from the new technologies that were developed, adapted and disseminated by the network of research and extension services that developed in the second half of the twentieth century has been confusing and often acrimonious. At one extreme, has been the argument that the rural poor was excluded from the benefits of this modernization (Griffin 1975; Hewitt 1976). This exclusion was deemed to be partly an effect of the ways in which technologies were generated. Public sector agricultural research services tended to conduct crop research on better quality land of the type that large farmers enjoyed but to which small farmers had less access; such research was also premised on an infrastructure capable of delivering adequate quantities of inputs (water, seed, equipment, agrochemicals) in a timely fashion and at the right price, and of marketing outputs. The inevitable result was that technologies generated were more likely to be appropriate to medium/large-scale farmers operating under favourable conditions and not to the complex and difficult conditions of small farmers. Likewise the researchers' focus on adapting high external-input technologies meant the technologies released were costly in cash terms. For resource-poor farmers these technologies were therefore either prohibitively expensive or at best, high risk because of the costs and delivery problems involved (Griffin 1975). Finally, the focus on technologies that replaced labour with capital was of less relevance to resource-poor farmers, and was particularly inappropriate for landless labourers who in effect lost out twice from this research focus: as they did not possess land, they could not make use of new technologies[10] and their jobs were displaced when large farmers adopted labour-saving technologies (Lipton and Longhurst 1989).

Other empirical evidence was however less damning of the new technologies, and has argued that, although the new technologies did initially favour larger farmers, over time more of the benefits have been captured by poorer groups (Dalrymple 1986; Rigg, 1989).[11] Increasingly it has been argued that distribution biases that emerged did not reflect problems with the technology *per se*, but rather with local institutional structures that led to social biases in access to the technology and to the related credit, information, and technical support services that would have helped resource poor farmers take advantage of these new technologies (Barsky 1990: 90; Byerlee 1987; Farmer 1977; Rigg 1989). In other cases, the style in which these services were made available was biased against small farmers (e.g. extension messages not using local languages; credit requiring land as a guarantee). It was, many have argued, these institutional support problems, not the technology, that resulted in an intensification of social differentiation and the concentration of wealth and resources in the countryside (Rigg 1989; Farmer 1977; Pearse 1980).

This more positive reappraisal of the much-heralded ills of the Green Revolution has drawn increasing attention to the ways in which farmers have themselves adapted these technologies. The implication, then, was not that technological modernization was entirely inappropriate to such small farmers, but rather that a more participatory approach to generating modern technologies would lead to new 'options' that farmers could put to better use (Rhoades and Booth 1982; Chambers *et al.* 1989). Indeed, in this more positive reading of Green Revolution strategies, some have gone so far as to argue that in certain institutional circumstances, the rural poor have been able to use agrochemicals and new varieties as a means of trying to survive the poverty-intensifying effects of wider forces of commercialization and land sub-division (Bebbington 1992; Lehmann 1984; Rigg 1989).

The implication of these more revisionist interpretations of agricultural modernization is that institutional changes could help recover and 'maintain the momentum of post Green Revolution agriculture' (Byerlee 1987), and spread the benefits of former and on-going research to the rural poor. This institutional structure would not simply have to bring extension services to areas that have so far lacked them, but also engage in styles of technical assistance that are appropriate for small farmers – methods such as the use of non-formal education and farmer paratechnicians. Likewise it would need to adapt its content and timing to the particular skills and needs most relevant to the circumstances within which the farmer was operating (Cotlear 1989). At the same time, the need would be for participatory institutional structures allowing farmers to make clear their needs and resources so that extension messages would be adapted to them.

For the rural poor with no reliable access to land the situation is potentially more complex. The emphasis of a technical support service relevant to their needs would have to be on developing skills increasing their strength in the rural labour market, generating labour absorbing technologies, and other income generating opportunities in rural areas requiring no more land than a backyard (such as poultry raising). Once again, however, technologies for the latter strategy could be developed and transferred through the sort of institutional structure being proposed.

Technological alternatives to the Green Revolution

Others are less sure about the validity of these revisionist arguments, and as the 1980s proceeded, the viability of the Green Revolution technical agenda was questioned with increasing cogency. Questions have been asked of it on ecological, economic and social grounds.

Use of chemical fertilizers and pesticides, for instance, has, it is argued, damaging effects on soil structure, causes groundwater and surface water pollution, and creates dependencies on external economic and supply systems. Similarly modern varieties and introduced crops are criticized for deepening dependencies on these inputs (an effect of breeding processes in which agrochemical inputs are used) and not exhibiting the qualities of resilience to a broad spectrum of local shocks that native crops and varieties typically possess.

Arguing against this high external-input package, agroecologists have insisted that sustainable production systems must be based on processes that occur naturally in ecosystems (Altieri 1987). Fertilization practices ought therefore be grounded in organic matter recycling, within systems that combine crop, livestock and trees; pests ought to be controlled by manipulating predator–prey relationships and natural pest repellents. Typically, agroecologists look to indigenous, pre-modern knowledge systems as a store of popular wisdom regarding such practices (Altieri 1987; 1990; Altieri and Hecht 1990).[12]

Such technical and ecological constraints to the sustainability of the modern option are aggravated by economic constraints which the crisis and adjustment programmes of the 1980s have made more apparent. Agroecologists have long pointed to the risks to small farmers implied by the monetary cost of agrochemicals. These risks have become more apparent as adjustment programmes have led to removal of some of the price distorting mechanisms that had kept nominal prices for agrochemicals and imported machinery below real prices *and* that at the same time had subsidized the credit used (more by larger farmers than small) to purchase these modern technologies. As this double subsidy is slowly removed, so the real cost of the agrochemical and fossil-fuel based mechanization package has become more apparent (de Janvry and Garcia 1992). This means that even if the continued power of commercial agricultural interests means that other price distortions are maintained and that therefore 'the agroecological alternative has yet to be given a fair market test' (de Janvry and Garcia 1992: 32), this situation is beginning to change. Increasingly the Green Revolution package will have to be assessed against agroecological alternatives without the hidden support of favourable price distortions.

This initially counter-cultural criticism of agricultural modernization has gained increasing credibility in orthodox institutions. Thus a recent ISNAR

18

review of challenges to agricultural research policy into the next century comments:

> Almost all the increases in farm output required to meet global needs . . . must come from further intensification of production practices on existing agricultural lands. . . . But productivity gains from conventional sources are likely to come in smaller increments than in the past and to an increasing degree. These gains are likely to be crop-, animal-, and location-specific. . . . These concerns mean that research must place greater emphasis on developing new farming methods and land uses that act to sustain or enhance the natural resource base for agriculture.
>
> (ISNAR 1992: 21, 23)

Even if such claims of conversion to the agroecological cause were designed to capture the initiative (Ruttan 1991: 401),[13] rather than being genuine commitments to sustainable development, they are significant.

Some suggest that the combined effect of these economic and ecological changes, and the realignments in institutions' priorities, has been to elicit the emergence of an alternative agenda for agricultural technology development (ATD)[14] based on sustainability concerns, and in the realization that it is no longer possible to maintain agricultural growth with capital inputs, and price distorting policies. It is also in some cases stimulated by the increasing urgency of reducing production costs in order to keep agriculture competitive in contexts of trade liberalization (FAO 1988: 16; IICA 1991; Kaimowitz 1991).

This agenda will be influenced by perspectives from agroecology. A focus on soil management using labour, land transformation techniques and organic material will be central, as will work on diversified farming systems based on natural ecosystem processes (Kaimowitz 1991). This in turn will mean applied agronomic research will become more important. Technological solutions will be location specific and so research will also have to be location specific (Ruttan 1991). Location specific adaptations in turn mean technologies will be information intensive rather than capital intensive. This will imply using more farmer knowledge (i.e. participatory ATD) but also providing support to farmers to increase their own knowledge and management skills.

Adapting and questioning agricultural modernization: institutional implications and the role for NGOs

While the emphasis of the agroecological agenda differs from the revisionist proposals for spreading the benefits of modernization to the resource-poor, there is much convergence. Both argue that by and large past national strategies of ATD have favoured larger farmers and have given little participation to the rural poor.

The two perspectives also converge around a set of common themes that future technological agendas should incorporate:

1 The generation of technologies more adapted to the types of agroecological conditions in which small farmers produce;
2 The strengthening of institutional capacity at a local level in order to enhance location specific ATD;
3 A widening of small farmer access to support services that facilitate use of technologies, and more equitable access to those services;
4 A reduction of agricultural production costs, to increase on-farm income and food security;
5 Wider availability of information among small farmers and an enhancement of their capacity to use it;
6 The conservation of the resource base in the long term, within the constraints of ensuring acceptable income flows in the short term;
7 Building an agricultural development strategy on the dynamism and knowledge of small farmers.

If we look at the past performance of the state and the market, their capacity, or interest, in meeting these challenges is clearly in doubt. Furthermore, as public sector research and extension budgets are cut back as part of the same neo-liberal policies, the public sector will be, and already is, even less able to sustain a local institutional presence.

Such doubts about the market and the state have drawn attention to the possible role of other not-for-profit, socially motivated local institutions. It is assumed that NGOs have a range of qualities that will allow them to (1) complement public sector activities, (2) press that public sector to be more responsive to small farmer needs, and (3) be important players in the new technological agenda (Altieri 1990; de Janvry and Garcia 1992; Kaimowitz 1991). Moreover, NGOs' institutional costs are not affected by public sector funding cutbacks.

Among other things, it has been suggested that NGOs are:

- More interested in low input agriculture than the public sector;
- More able to adapt technologies to local conditions than the public sector;
- More participatory than the public sector, and so more able to link local and modern agricultural knowledge;
- More efficient in their own use of resources than the public sector;
- More oriented towards systems perspectives on ATD than the public sector;
- Aware of and able to represent the aspirations of the rural poor.

We will address these assertions in Chapter 3.

THREE BROAD MOTIVATIONS IN THE INTEREST IN NGOs

Many roads may lead to NGOs – but we can perhaps identify three main motivations behind this interest. All three originate from different questions that have been increasingly asked about orthodox strategies and models of development.

Restructuring the management of modernization

For some, the interest in NGOs has been stimulated primarily by a search for new and more efficient ways of achieving relatively orthodox goals of development. The interest in NGOs as new extension agencies to continue distributing 'modern' technologies is perhaps the best example of this. In general terms, the hope that NGOs might be able to do what donors and policy makers had expected of the state reflects a rethinking of the institutions best able to implement development – but no great rethinking of what that development should look like.

Alternative development: governance, empowerment, human agency

A second road to NGOs has been inspired by deeper questions asked of existing models. There is a certain convergence between the concerns of those who talk of good governance and democracy, those who talk of empowering the poor, and those who stress the need to respect the poor as human agents who can, and should be allowed to take active roles in fashioning the contexts in which they live. All are elaborating an argument that development should be more participatory, diverse and inclusive than in the past. All are questioning the dominance of economistic thinking in definitions of development. All have argued that for future development to be reoriented so as to respect these concerns, more spheres of local action and support should be passed over to non-governmental institutions, and that the role of NGOs ought to be strengthened.

Greening development: NGOs and the environmental imperative

Finally, perhaps the most fervent interrogation of development experiences to date has come from the environmental lobby, and those who insist that development has undervalued the environment, been subsidized by all manner of unpaid-for externalities, and operated without a view to the long-term sustainability of both place and planet. These perspectives, coming from both conservative and more radical agencies, have seen in NGOs a force for a greener development. They have pointed to the important role that NGOs, both indigenous and international, have had in the promotion of sustainable development and agroecology. They have therefore argued that increasing the scope of NGO action will help widen the impact of agroecological perspectives on actual development interventions (Altieri 1990; Farnworth 1991). At the same time it has been suggested that the lobbying power of these NGOs is an additional force for greener policies.

21

In some regards the growing interest in sustainable development, and the tendency to broaden its definition beyond narrowly environmental terms, cuts across and combines these three themes (Goodman and Redclift 1991). It represents the hope of a development that can empower, sustain the environment, combine new and old technologies in benevolent ways and respect diversity. The close association of NGOs with sustainable development (which even reached our television screens during the United Nations Conference on Environment and Development at Rio in 1992) has been an extremely powerful force pushing NGOs to the fore of development debates.

QUESTIONING NGOs: WHAT ROLES FOR THE STATE?

In the foregoing discussion we have argued that there has been a range of currents in contemporary debates on development that have furthered the NGO cause. Indeed, in their more jubilant moments these discussions have fostered positive, at times mythical, thinking about the good that NGOs can do.

However, as we have also signposted, there are many questions that are being asked of NGOs. Indeed, in some quarters, we might already be able to detect a certain disenchantment. Part of this stems from the sorts of criticisms of NGOs discussed earlier, and part from assessments of the limitations of the NGO model of implementing development.

Issues of poor NGO co-ordination and communication

Among the most frequently heard criticisms of NGOs is that they are not subject to mechanisms of accountability or co-ordination. Thus they receive money in the name of the rural poor, but the only mechanism to ensure that the funds are properly used is follow-up by the financing agency. To developing country governments this is doubly problematic: it neither permits them to monitor financial propriety and impact, nor does it allow them to ensure that the work of NGOs is in accord with government policy. Of course many NGOs are deliberately in disagreement with government policy and so would not wish to be co-ordinated. On the other hand, by acting independently the NGO can in fact weaken government efforts by undermining their legitimacy, or contradicting their approaches. The difficult decision for financing agencies is to know when the NGO should be expected to co-ordinate with government programmes, and when not. This is clearly a political decision. Questions of co-ordination are further discussed, and examined against empirical evidence, in Chapter 5.

Further twists to this problem weaken the NGOs' own contribution: distance from government programmes and policy making may protect NGOs' independence – but it also means they remain dependent on a wider, often foreign, policy environment over which they have no influence. This is a policy environment that frequently determines the outcome of NGOs' projects (Bebbington 1991b; Carroll 1992; Clark 1991). Secondly, retaining independence from any co-ordinating mechanism often leads to situations where within one region several organizations (NGOs and public) are each merrily working away duplicating efforts, replicating mistakes, using conflicting approaches, and generally confusing the rural poor. Such circumstances may sometimes arise because of deliberate neglect of each other – because NGOs end up competing for clients, and against each other (Ayers 1992; Kohl 1991). Either way, it weakens and indeed may negate the contributions of NGOs. Some of the consequences of weak inter-NGO co-operation are discussed in Chapter 5.

Issues around NGO size and scaling up

A second set of observations revolve around the problem of NGOs' small size and limited impact. While smallness may be an advantage, it can also bring limitations. In this regard Sheldon Annis's (1988: 209) quotation about NGOs in general has been frequently stated, but merits repeating here: he comments:

- In the face of pervasive poverty ... 'small-scale' can merely mean 'insignificant';
- 'Politically independent' can mean 'powerless' or 'disconnected';
- 'Low-cost' can mean 'underfinanced' or 'poor quality';
- 'Innovative' can mean simply 'temporary' or 'unsustainable'.

These were the observations that led Annis to early discussions of 'scaling up' – or, how to increase and sustain NGO impact. Interestingly Annis suggested that in fact the strongest NGO impacts were in areas where government presence was significant and where there was some degree of NGO contact with government. In a recent synthesis of scaling up experiences, Edwards and Hulme (1992b: 78) have argued there are perhaps four approaches to scaling up: (1) working with government to spread NGO methods and change policy, (2) lobbying and advocacy, (3) operational expansion, and (4) strengthening networks of base organizations. The first two involve interaction with government, and the latter assumes that base groups will subsequently negotiate with government. Chapter 5 explores the empirical material collected for this study against some of these approaches.

Again, in the scaling up debate the conclusions can cut both ways. As Munck comments:

The central dilemma of alternative development is that its initial emphasis on smallness must be increasingly framed in terms of a national and even global strategy, without this concern for larger issues stamping out the autonomy of local action.

(Munck 1992: 178)

Faced with this dilemma, some suggest that to scale up can do the NGO more harm than good, making it too big or too politically compromised. Others suggest that ultimately NGOs have no option: if they do not try to increase their impact through policy reform and multiplication of their models, then they will ultimately be meaningless – a few social development straws finally to be blown away by a neo-liberal tempest (Carroll 1992).

Whatever the resolution of the scaling up debate for any particular NGO, it has placed the issue of interaction with the state at centre stage.

Issues around NGO capacity

A further problem with NGOs' smallness is that it limits the technical and professional resources at their immediate disposal. In the agricultural sphere, this means few NGOs have technical libraries, research station and laboratory facilities, and all have limited capacity to do controlled research. They lack time, personnel and money. Similarly, they are unable to have many different specialists within the same organization, and so it is frequently the case that a technical problem arises for which the NGO has no internal capacity to respond.

These constraints are likely to become much more severe as peasant agriculture becomes increasingly subject to competitive pressures unleashed by free trade. For while NGOs may have more experience of low input technologies than does the state, it is less clear they have all the skills required to identify the technological and marketing strategies necessary to respond to these new challenges.

These limitations raise the question of where NGOs might look for such expertise to complement their own capabilities. Sources that they have explored include universities in the North and the South. Additionally, the formation of NGO networks to share knowledge and resources is a step towards trying to facilitate such mutual professional and technical support. None the less, despite the crises that the public sector's research institutions are experiencing, in many there remain specialist human and infrastructural resources which NGOs could draw on to compensate for the gaps in their own skill bases.

Issues around 'representativeness'

As we have suggested, while NGOs may be seen by some as vehicles of democratization, their own house is not always in order in this regard. Many do not have the rural poor represented in their directorates and ultimately control resources and decisions from within. There are mechanisms of accountability between them and the rural poor but these are mainly informal, or depend on the stringency of their donors' requirements. In cases where governments have moved back to electoral democracy some have suggested

24

that one means of holding NGOs more accountable would be through some form of link with the government. Not surprisingly, governments have been among the first to say this. However, we must not be naive – governments are often more motivated by the desire to gain access to NGO funds, or monitor NGOs they fear as political competitors, than by a desire to make NGOs accountable to the rural poor. This ulterior motivation frequently leads NGOs, and their donors, to fight shy of any government monitoring. Thus, in Bolivia in 1990, the government passed a decree for NGO registration that included clauses for a tax, and for the representation of the cognate government agency in all NGO evaluations in the future. The stated rationale for this was that government needed to know of NGO innovations, and that NGOs ought be more efficiently co-ordinated with government. But many in the NGO and donor community in Bolivia felt the prime motivation was a desire on the government's behalf to gain access to NGO funds and to monitor the activities of organizations sympathetic to the opposition parties.

CONCLUSION AND SUMMARY

Much of the enthusiasm for NGOs has been motivated by crises in the state's performance over the last several decades. State inefficiency in programme management has now been compounded by economic crises which leave public agencies even shorter of funds, and even more inefficient. At the same time, the concentration of resources in the public sector has led to the colonization of the state by interest groups. Meanwhile, these states have frequently repressed democratic process.

Such limitations in the state have long been criticized from radical and social quarters. Now, it seems, international financial institutions influence governments in forcing these to accept policies that aim to restrict the size of the public sector. These institutions have promoted neo-liberal policies and also, with growing vigour, the idea that governments should cede development interventions to NGOs, at the same time creating more chance that these NGOs will be effective and successful in areas where the government was not. Concern for sustainable development has similarly turned their attention to NGOs. At the same time, debates in academic circles have also begun to take NGOs more seriously. From many sides, interest in NGOs has grown rapidly.

Yet there has also been recognition of some of the limitations of NGOs: small size, restricted impact, distance from policy decisions, professional and technical limitations, poor co-ordination, problems of representativeness and accountability. Discussion of these failings has ironically led the discussion back to the state – for in different ways it has been suggested that some of these limitations might be addressed if NGOs engage in more co-ordinated relationships with public sector organizations.

The remainder of this book addresses these strengths and limitations in NGOs' contributions to development, and then looks more specifically at the issue of their interaction with government. However, as we do so, we are not trying to make generalized statements about what NGOs are and what their strengths and weaknesses are; nor are we offering recipes for how to organize NGO–state relations. What we are doing instead is to point to the range of questions that must be asked of situations involving NGOs and their relationships with the state. While, then, we are willing to generalize about questions, we are much more guarded in generalizing about explanations. Consequently, readers will find that the case studies we discuss do not fall neatly into particular boxes – rather they are drawn on selectively, and sometimes repeatedly, to illustrate different points.

Starting from the context of the evolution of NGOs and the state outlined here, and of interactions between them, Chapter 2 examines in more depth the study's initial hypotheses concerning the scope for mutual complementarity between the two sides in the functions of agricultural research and extension. It then examines how scope for exploiting this functional complementarity is influenced by socio-political dimensions of the NGO–state relationship. Chapter 3 then considers NGOs' work in technology development, assessing whether this reflects the claims that have been made for NGOs, and making certain NGO strengths and weaknesses in agricultural development more explicit. The fourth chapter does the same for the question of NGOs' relationship to the rural poor – both in terms of their capacity to work with the poorest of the poor, and in terms of their approaches to participatory rural development.

In Chapter 5 the discussion focuses specifically on the question of state–NGO interaction, and assesses how it has worked in the past and whether it has had, or in the future will have, a positive impact on NGO limitations. Our conclusion is that, although highly diverse, on balance such interactions and linkages can have positive implications for the quality of development work done with the rural poor. However, the relationship is an uneasy one, and the final chapter considers in more detail the factors that have most impact on whether such relationships will be possible, and fruitful on a wider scale for the future. In that chapter our conclusions are more guarded. There remain many obstacles to successful NGO–state relationships, which must be borne in mind in any development intervention that plans to bring together or work with, both sets of actors. Donors, especially, must remember that as in the past so in the future, NGOs and the state will remain reluctant partners.

NOTES

1 In 1991, US$7 billion were channelled through NGOs to the developing world, and most official aid agencies channel 10 per cent or more of government aid through NGOs (Riddell and Robinson 1993).

2 The genealogy of the concept of the 'third sector' is complex: Korten (1987), for instance, obtained the concept in good measure from Esman and Uphoff (1984),

whose thinking had been stimulated in part by earlier work at ODI (Hunter 1982). The concept has also been elaborated by Uphoff (1986; 1993).

3 Another label is 'private voluntary organization' or PVO. This is a term used frequently in the USA referring to a very broad range of non-state organizations. Its usage is quite similar to this very broad use of the term NGO.

4 These include, for instance, CRS in Gambia, Rodale in Senegal, and CARE in Kenya.

5 Again, the distinction between membership and non-membership organizations has been around for some time: Carroll acknowledges the concept to Uphoff (1986), which, in turn, drew upon Esman and Uphoff (1984).

6 The discussion in this section is not meant to suggest that donors have ceased to support government implemented extension, nor that all are rushing to involve NGOs in their projects. We are simply pointing to a trend that seems to be occurring and is creating more interest in NGOs.

7 See for instance, the discussion of private research and development organizations in Nepal by Shrestha and Farrington (1993).

8 We use 'Green Revolution' as shorthand for a set of agricultural technologies which broadly comprise: highly fertilizer-responsive crop varieties which (especially in the cases of rice and wheat) were resistant to lodging, but generally required higher crop protection inputs and, partly because of their shorter growing period and the higher cropping intensities that this permitted, also required more closely controllable water supply and, often, higher levels of mechanization than did traditional varieties.

Some have argued that some Green Revolution strategies were premised on the measurable transformation of the traditional into the modern through the transfer and diffusion of technology and institutional models in ways which were highly centralizing and which disregarded the capabilities of existing indigenous R&D systems, whether formal or informal, and the local social and cultural contexts into which technologies were to be transferred (Biggs and Farrington 1991).

We recognize, however, that many of the components of orthodox modernizing approaches – high external-input varieties requiring high levels of infrastructural and institutional support, and the insensitive transfer of these into local settings – have a much longer history than the Green Revolution.

We also recognize that there is a long history of scientific effort, focusing especially on the manipulation of genetic material, which is of high potential benefit to the rural poor who seek livelihoods in complex, diverse and risk-prone farming conditions. This includes work on pest and disease resistance and tolerance to adverse soil and climatic conditions. Similarly, some institutes originally in the vanguard of the Green Revolution have for some years had major programmes in these areas, and, more recently, on issues of wide agroecological concern, such as the long-term sustainability of irrigated rice systems, and the impact of salinization.

9 This discussion draws heavily on the discussion in the third chapter of Bebbington and Thiele (1993).

10 It should be noted that landless labourers may be able to use sharecropping contracts as a means of gaining access to land, and as a result be able to use these technologies (Lehmann 1982; 1984).

11 These groups may also have benefited as food consumers, by way of cheaper food (ISNAR 1992).

12 There is perhaps a correlation between the rise of agroecology and the crisis of left-thinking: one form of critique has been replaced by another, and the object of critique has shifted from capitalism to modernity.

13 Ruttan was referring to the new found interest of the US National Research Council in alternative agriculture.
14 Agricultural technology development (ATD) is defined to include the processes of research and dissemination. Furthermore, we use it to include the development of technology by farmers and its transfer from them to extension agents, as well as vice versa (Wellard and Copestake 1993).

2

CONCEPTS FOR ANALYSING
NGO–STATE RELATIONSHIPS

INTRODUCTION

With the growth in interest in NGOs among policy makers and social development researchers, it is necessary to develop the conceptual frameworks for analysing these organizations. On the basis of such frameworks hypotheses can be developed about how NGOs operate and thus about the nature of policy possibilities for working with NGOs. In particular, the interest in NGOs' linkages with the state requires that decision makers in government, donor and NGO circles think carefully about the terms on which this relationship can be structured.

Some authors have initiated this type of conceptual work. NGOs have been discussed, for instance, within the relationship between civil society, state and market (Fowler 1990; Fox 1990a; Uphoff 1993), as parts of civil society (Clark 1991; Durán 1990), in the terminology of organization theory (Uphoff 1986; Esman and Uphoff 1984), and in the context of the research–extension continuum (Martínez-Norgueira 1990). These are all starting points, but it is symptomatic of NGOs' recent rise to prominence that the conceptual tools (and empirical information) for understanding them as sociological phenomena are still at an early stage of their development.

In this chapter we present a selective review of approaches to understanding how NGOs operate as one of a number of actors in the development of agricultural technology. The main intention of this discussion is to elaborate on those frameworks that have proved particularly useful in this study, and that generate hypotheses that are relevant to a reading of the empirical chapters that follow. A subsidiary intention is to point to areas in which these approaches can be usefully combined to provide a link between theoretical and policy frameworks, and to provide entry points to decision makers.

Our discussion revolves around two very broad approaches to organizations in rural and agricultural development. The first is a perspective that focuses on the comparative advantages of different organizations in fulfilling specific functions in agricultural technology development. In general, this is the most explicitly policy and management oriented body of literature, and indeed much of it has been developed at agricultural research centres, development consultancies and research projects that have had close contacts with donor agencies. The second perspective is more closely associated with academic research traditions. In part it focuses on the social origins and histories of institutions, tracing their emergence, and position in the wider relationships between social groups. In some versions, such approaches have had a strongly structural and political economic feel to them, and in the case of NGOs have often analysed them as necessarily opposed to the interests of state and capital. In recent years, the deterministic tendencies in this tradition have been challenged by discussions that revolve around concepts of civil society and state, linking structuralist approaches with those discussions that pursue a more 'actor-oriented' approach (Booth 1992). We suggest that it is in these actor-oriented approaches that the discussions of NGOs can be made policy relevant at the same time as they are tied into conceptual analyses of institutions and rural development (cf. Booth 1992).

FUNCTIONAL ROLES AND INSTITUTIONAL ADVANTAGES IN AGRICULTURAL TECHNOLOGY DEVELOPMENT

Functional approaches to ATD

Policy oriented discussions of ATD are frequently couched in very functional, or task-oriented, terms. Analyses of the relationships between research and extension are a simple example of studies using this functional terminology. A conceptual framework that focuses on functions and tasks in this way is often particularly relevant and accessible to decision makers such as research managers. By building a framework on activities (or what may be called lay concepts) that constitute part of the everyday world of those managers, it makes it easier for them to see the direct practical implication of a conceptual argument to their own work. For instance, a conceptually grounded analysis which concludes that ATD in a particular country is undermined by problems in the research system, immediately leads a decision maker to think of the research institutions where changes must be made.

Among these approaches, of particular relevance to this study are two large research programmes at ISNAR that have grappled with the problems of institutionalizing client participation in agricultural research, and improving the relationships between agricultural research and extension (Merrill-Sands and Kaimowitz 1989).[1] Although these studies had a largely public sector focus, their concepts are, with amendments, relevant to the discussion of NGOs.

In one of these studies, Kaimowitz et al. (1990: 231) talk of the *Agricultural Technology System* (ATS), defined as comprising all the individuals and groups working on the development, diffusion and use of new and existing technologies, and the relations between, and actions of, these individuals and groups. These agents perform different functions which constitute *sub-systems* in this ATS. Each of these sub-systems carries out a specific *function* in the ATD process and each one potentially involves a group of actors. The *basic research sub-system* develops new knowledge, and the *applied research sub-system* develops new technologies on the basis of this knowledge to tackle certain concrete agricultural problems. The *adaptive research sub-system* effects changes in the technologies to adapt them more to agro-ecological conditions and to the socio-economic realities of specific regions and producer groups. The *technology transfer sub-system* facilitates the adoption of technologies by the users. Finally, we might also refer to a *sub-system of technology use*, within which technological practices are implemented at the level of the peasant production unit. In all these sub-systems, the relevant actors are not only researchers and extensionists, but also farmers – because the general development of agricultural technology includes learning from farmers as well as the institutional generation of technology to be transferred to them (Wellard and Copestake 1993).

The ISNAR studies then looked for mechanisms of inter-institutional co-ordination and collaboration that would improve the effectiveness and efficiency of linkages between actors and institutions in the different sub-systems. It was argued that ATD would be more effective and relevant to small farmers if the links between the sub-systems are strong and allow effective flows of information. In particular, the studies suggested, mechanisms had to be found (1) to increase the involvement of small farmers in research and extension activities, and (2) to improve the communication channels between extensionists and researchers so that extensionists might be able to feed information on farmer conditions and preferences back to research and influence the type of research done, and to select technological options more effectively from the technologies researchers have already developed. As the ISNAR studies also suggested, however, improving such linkages within the public sector is fraught with difficulties.

Functional demands and shortcomings in government capacity

These types of discussion lead to questions about what type of organization and organizational structure are best able to perform the different functions outlined, and the linkages between sub-systems. Traditionally these questions were answered in public sector terms, and as we noted the ISNAR studies had a strong public sector focus. When the current structure of the public sector national agricultural research and extension service (or NARS)[2] was not performing these functions adequately, the response was to look for appropriate restructuring, or to create new departments to be responsible for unfulfilled functions. The clearest cases of this are the attempts to create special programmes for on-farm research, for farming systems and socio-economic

31

research, and for research–extension liaison (e.g. Soliz *et al.* 1989; Kean and Singogo 1990).

Over time, these analyses of the functional capacities of government institutions have become increasingly sanguine about the capabilities of government. Political interference to use government services for purposes of patronage, or simply to orient NARS activities to the needs of medium and large agro-export and commercial farmers, has been one of several recurrent problems. Similarly, bureaucratic inertia and inflexibility can undermine the ethos of service provision in the NARS: budgetary restrictions (which are particularly severe under structural adjustment programmes) and delays in central government agreement to and disbursement of departmental budgets all impact most severely on operational expenditures, given that payment of salaries has the first claim on budgets, and that pressures for staff cuts are generally resisted.

Among the factors constraining the orientation of public sector research and extension services towards farmers' needs, and their development and dissemination of appropriate technologies we might draw attention to the following:

1 Hierarchical and centralized management structures;
2 Inadequate scientific management;
3 The absence of performance-related reward systems;
4 Inadequate operational budgets;
5 Institutional barriers to communication and to joint activities between research and extension;
6 Inappropriate research methods and gender biases.

Hierarchical and centralized management structures

Many of the prevailing structural difficulties in government research and extension services originate in institutional forms introduced during the colonial period – often for research on export crops of particular interest to the colonial power. These structures are typically organized by commodity and/or discipline. They tend to be strongly hierarchical.

Decisions on even the most routine issues – such as the ordering of spare parts (especially those requiring foreign exchange), tend to be taken at or near the apex of the hierarchy. While this concentration of decision making did not pose particular problems if most staff were located at one site[3] delays become more severe once a national network of stations, sub-stations and experimental farms is set up to deal with the production of diverse food crops in different regions of a country.

In Nepal, for instance, the government's own Socioeconomics Research and Extension Division, reporting on the performance of government research farms and stations, catalogues a number of problems in part attributable to centralized decision-taking. These include: delays in the release of annual budgets from central headquarters; delayed approval for repairs to equipment and buildings; and delays in the authorization of recruitment to vacant posts. They also note that, driven by administrative necessity, much of the communication is between field stations and headquarters, at the expense of horizontal communication *among* research stations, so that gaps and overlaps in coverage are rife (see, for instance, Sharma *et al.* 1988, summarized in Farrington and Mathema 1991).

Inadequate scientific management

Much recent 'farmer first' debate has been initiated by those (usually social scientists) who argue that it is the 'reductionist methods' of normal science that have given rise to top-down approaches to transferring technologies which have little prospect of successful adoption because farmers' requirements are not adequately taken into account in technology design. While there is some truth in these arguments, they miss the more fundamental point that quality control in many public sector research organizations is poor. Evidence of this is found, first, in the widespread duplication of effort among research projects, especially in large systems (Coulter and Farrington 1988); second, in the inadequate effort to locate new project proposals into the context of previous work, within or outside the country, and, third, in neglect of arrangements for monitoring the progress of research or evaluating its impact, so that little effort is made to obtain feedback from users. Part of the problem lies in the inadequate design and functioning of bibliographic and management information systems, but there is evidence that, even where bibliographic services are good, they remain inadequately used (Ferguson, pers. comm. on Malawi), inaccessible, or used predominantly by research students, not by scientists (Hans, pers. comm. on India).

The absence of performance-related reward systems

Many research managers are locked in a vicious circle of poor information about the progress and outcome of existing projects, and the absence of criteria for the selection of new research, leading to inability to monitor the performance of scientists, which, in turn, contributes to the absence of performance-related reward systems. Where measures of performance do exist, they rely heavily on quantitative assessments of published output. This criterion is known to distort research agenda away from innovative efforts, whether in the field, on station or in the laboratory, and towards research of a more routine kind practically guaranteed to provide publishable outputs. It also biases research towards those issues which editors of in-house or national journals regard as publishable, such as work involving new techniques. Clients' needs receive scant consideration in such cases.

Responding to this problem can be further complicated in those cases (as in

33

S. Asia)[4] where a high proportion of the total value of emoluments is made up by long-term benefits (pensions, housing, medical benefits, children's education). In such situations, the marginal impact of performance related pay is frequently limited.

Inadequate operational budgets

Where budgets are cut – as has happened in many countries under economic reform programmes – resistance to staff redundancies has led to a disproportionate reduction in operating expenses (Pardey et al. 1991). This – often accompanied by delayed disbursements from central government – has made it particularly difficult to plan and implement field projects. Biggs (1989: 31) notes that 'activities that encourage farmer participation in the research process are often the first to be cut back in periods of austerity'. As one of numerous examples that support this statement, Galt and Mathema (1987) note that the participatory and rapid diagnosis technique developed in Nepal (*Samahik Bhraman* – a trek involving scientists of different disciplines) has in practice been constrained by the fact that daily travel and subsistence allowances do not cover the costs necessarily incurred by scientists.[5]

Institutional barriers within research systems, and between research and extension

Farmers, particularly in difficult areas, generate livelihoods from a wide range of farm and non-farm activities (see for example, Jodha 1986). Conventional disciplinary or commodity-based approaches to research are therefore unlikely to meet their needs, and widespread efforts have been made to introduce systems-based interdisciplinary approaches. However, the progress in institutionalizing this has been slow (e.g. Collinson 1988).[6] Others have attempted to make interdisciplinary approaches less threatening institutionally by organizing temporary 'task forces' (e.g. at WARDA). However, despite these efforts, commodities and disciplines remain the dominant dimensions of research organization in many countries, and potential responsiveness to farmers' needs remains unrealized.

Furthermore, the widespread perception among scientists that their status is superior to that of extension workers, together with the fact that extension departments are commonly housed in separate buildings – often reporting to different departmental heads – results in weak links between research and extension, which are exacerbated by the lack of interest of some scientists in feedback on clients' response to the technologies deriving from their research.

Research methods

The research needs of plantations producing export commodities, or of large farms operating under favourable agro-climatic conditions which produce a narrow range of commodities can usually be met adequately by research which is dominated by discipline- or commodity-based approaches. Furthermore, these relatively simple farming systems can easily be replicated on research stations and, once produced, new technologies can be transferred from stations to the entire client group.

The relevance of this approach to small farmers was increasingly questioned in the 1960s and 1970s, when the rationality of small-farm systems incorporating a wide range of enterprises (crops, trees, animals), combining crops into mixtures and relays, and drawing heavily on off-farm resources for fodder and green manure became firmly established through the pioneering work of Norman (1974) in W. Africa, Collinson (1972) in E. Africa and Mellor (1974) in India. This work stimulated calls for new approaches to agricultural research which would be systems-based, interdisciplinary and conducted largely on-farm. They were accompanied by growing confidence that rapid appraisal methods could satisfactorily replace much of the time-consuming (and frequently inaccurate and irrelevant) collection of diagnostic data by questionnaire-type surveys or detailed farm records (Collinson 1972; Longhurst 1981). This, in turn, led to the call for approaches which would take into account more fully than hitherto farmers' perceptions of needs and opportunities, as well as their local knowledge (Chambers and Ghildyal 1985; Chambers et al. 1989; Farrington and Martin 1988).

However, rhetoric of this kind is far ahead of reality in government services (and many non-governmental organizations as well). The problems we have already listed constitute important obstacles to these methodological innovations. In large measure this is because such innovations have corollary implications for institutional change. They imply the need for greater decentralization, more farmer participation in research and resource allocation decisions, more work on farm, and a move away from commodity defined departments to a structure based on farming systems.

These problems are accentuated in those countries that are small in an economic and 'demographic' sense:[7]

1 In relation to their overall population size, expenditures on agricultural research in small countries are no less – and in some cases higher – than those of large countries, but levels of agro-ecological and socio-economic diversity do not diminish in direct proportion to population size or area. Small countries therefore have disproportionately high levels of diversity to deal with;

2 Conventional types of experimentation, involving replication and multi-site testing, require a critical minimum mass of effort, and certain facilities (laboratories, libraries) and equipment are 'lumpy' assets, and so disproportionately expensive for small countries;

3 Small systems are less able than large to offer possibilities for transfer and

advancement; promotion posts are likely to be few, and may be 'blocked' by their incumbents for long periods;

4 For many of the above reasons, staff turnover tends to be high. This is less of a problem where staff join other organizations within the country (such as NGOs) and continue to work on agricultural development issues. It is a more severe problem where they join international organizations. But, even in the former case, levels of attrition on the research service can quickly become so high as to make it unviable;

5 International networks – whether for collaborative research or for information exchange – have much to offer small countries, but at the same time, requirements such as those to participate in regional collaborative trials, place disproportionately heavy demands on them, as does attendance at regional-level meetings.

This litany of institutional constraints on research and extension capacity in the public sector can be overwhelming (and we could have listed more). Viewed in the context of the ATS as outlined above, they imply above all that the structure, culture and rules of many government organizations are such that they are severely constrained in their ability to perform the tasks of on-farm, participatory and adaptive research, and of communication of information on farmer preferences, needs and ideas between the different sub-systems of the ATS. It is also painfully clear that this capacity is being further undermined by the budgetary constraints arising from cutbacks in public sector expenditure.

These observations, and contemporary economic realities, demand that careful thought be given to the tasks that the NARS can best achieve with the limited financial resources at their disposal. These questions cannot however be answered without corollary consideration of which other institutions would have the capacity to assume responsibility for the tasks in certain sub-systems of the ATS from which the public sector might withdraw in any such strategic targeting of state resources.

Non-governmental capacity to assume functions in the ATS

Parallel to these observations of the public sector, the last decade has seen a growing interest in the special capacities of non-governmental organizations. An emerging literature on NGOs has begun to suggest that, as befits the generic descriptor 'non-governmental', NGOs were everything that government was not. Where government services were exclusionary, commodity based and high input, NGOs were participatory, systems focused and low input. Much of this literature has had a strongly romantic turn, but somewhere between romance and cynicism there remains an element of truth.

This literature contrasts the typical structures of the public sector with the shallow hierarchies of NGOs, claiming that the speed with which decisions can be made among NGOs strengthens their capacity to respond to clients' needs. For instance, Esman and Uphoff in 1984 suggested that what they called 'private voluntary organizations' (PVOs) had certain important advantages over government procedures in development administration because of their commitment, patience and capacity for experimentation (Esman and Uphoff 1984: 274–5). The articles edited by Drabek (1987) made similar claims. One recurrent theme has been that their institutional structure gives them an advantage in responding to the needs of the rural poor: 'the effectiveness of NGOs emanates mainly from their small scale, their flexibility and their project orientation; . . . their legitimacy at the grassroots level is well established' (Martínez Norguiera 1990: 99). As against government, which has to serve a heterogeneous clientele and has a complex structure, Martínez Norguiera (1990) suggests that NGOs have the comparative advantages of having a more homogeneous clientele, and a more clearly defined internal structure.

It is, perhaps, NGOs' small size and internal flexibility that commentators have emphasized as among their strongest institutional advantages. These characteristics can facilitate creativity within the organization, because they generally present fewer obstacles to change than in the large and much more inflexible NARS. Similarly, the work environment and initial recruitment decisions often lead to a motivated and socially committed staff. This motivation is also strengthened by reward structures, and wage levels that are often higher than in the public sector.

These institutional qualities give rise to certain functional strengths. For instance, as we discuss in Chapter 3, NGOs have often been quicker to experiment with different farming and food systems approaches to ATD, and with participatory approaches to research and extension (Farrington and Martin 1988). Many of the cases of technology development based on the principles of agroecology and indigenous technology come from NGOs (Altieri 1990), as do many of the cases where farmer extension agents have been trained.

Similarly there has been greater flexibility and innovation with concepts of farmer participation in NGOs, many going beyond participation in on-farm research to the strengthening and at times creation of local organizations. Such organizations increase the possibility that after the NGO has moved on, local groups will be able to exercise a demand pull on other institutions. Smallness of size can also facilitate rapid response to new demands from local clients and to technical problems as they arise. More generally, as noted already, this small size and local orientation can allow the institutions to respond to the challenge of adapting technical support to particular circumstances more effectively than can the centralized structures of NARS.

Any enthusiasm for NGOs' institutional strengths and transformative potential needs, however, to be tempered. David Lehmann (1990) referring to this enthusiasm as *basismo*, offers an early contribution to this more realistic vision. Lehmann uses the term *basismo* to refer to that complex of populist ideas embracing a faith in the capabilities of the rural poor, in the potential of local action and the merits of sustainable development, and a hope for a more direct and participatory democracy. In an essay he mischievously titles '*Basismo* as if reality really mattered' (1990: 190–214), he endorses the spirit of this *basismo*, but argues strenuously that it will come to little as long as it, and the NGO movement which he feels is particularly instrumental in fostering this populism, remain guided more by will than skill, hope than strategy and an interest in local empowerment rather than policy change. Lehmann urges NGOs and other *basistas* to gird up their loins and take seriously the challenge of changing the state, indeed as the most serious challenge of all. Building a more transparent, honest and efficient state is, for him, the *sine qua non* of a *basista* strategy whose ultimate goal must be broad based change through policy and institutional change. Local groups can play an instrumental part in this 'modernisation from below' (ibid: 190), but they will only do so if they engage with state.

As Lehmann argues, NGOs and grassroots groups are plagued by many weaknesses that will confound policy change if they only act alone, and locally. We have already rehearsed these points in Chapter 1. Here it should simply be emphasized that many of these weaknesses in NGOs stem from their small size, institutional structure and donor dependence. In short, the structure of NGOs seems to suit them best for the more local functions in the ATS, and that of the NARS for the more centralized functions. If only we could wish away politics, then the effectiveness of both would be enhanced by improving the links between them, and so between the different tasks that they best perform – if only.

Tying the knot – or an arranged marriage? Seeking functional complementarities between NGOs and government

Discussing functional complementarities in these ways has inevitably led to proposals that functional responsibilities in the ATS should be divided among NGOs and the public sector according to their strengths and weaknesses. Broadly speaking, NGOs are held to be strong in identifying clients' needs, taking rapid decisions on how to respond to needs, and supporting local initiatives. Their weaknesses lie in their small scale, and in fragmented initiatives, poorly co-ordinated among themselves and with wider develop-ment processes, and at times driven by an excessive preoccupation with demonstrable action. Government has a potentially complementary set of advantages in that it controls major policy instruments, possesses a broad revenue base, and has (more than NGOs at least) capacity to undertake large-scale infrastructural investment and address complex technical issues. It also has the complementary disadvantages in cumbersome decision-taking and interventions inadequately designed to address clients' needs.

Table 2.1 Functional strengths and weaknesses of NGOs and government in agricultural technology development for the rural poor

	Government	NGOs
Strengths	Specialist facilities (laboratories, libraries) too 'lumpy' for NGOs to own Specialist skills in disciplines (entomology, plant breeding, etc.) and in experimental design and analysis Links with wider scientific community both national and international	Participatory methods deriving from an agenda for technological change based on farmers' needs Holistic approaches integrating technology change with nutrition, education, marketing, processing, etc. and addressing constraints arising in these spheres Concern to create a demand-pull on government institutions, and to support the emergence of local groups capable of this and of some agricultural experimentation Research tends to be 'issue'- and 'action'-oriented; research–dissemination–feedback seen as a continuum
Weaknesses	Client-orientation, where it exists, tends to be towards powerful interest groups Limited awareness of and responsiveness to the needs of the rural poor Research methods inappropriate to the rural poor: inadequate diagnosis; excessive on-station research; non-participatory approaches Discontinuities between research and dissemination Poor monitoring and feedback on technologies developed Inadequate co-ordination among government departments facilitating technological change	Limited capacity for research; limited technical skills Weak links with wider scientific system Small scale efforts; often poor institutional memories; weak links with other NGOs working on similar technology issues

Indeed, when presented with summary tables such as Table 2.1 it is difficult to resist making proposals for increasing NGO responsibility for activities in the adaptive and transfer sub-systems, and co-ordinating these activities with work done by government institutes and international agricultural research centres (IARCs) in the applied and basic sub-systems.

Such an allocation of tasks, it might be suggested, would, along with closer linkages with NGOs, bring to government research and extension services a number of advantages including:

1 A stronger awareness of and orientation towards the needs of the rural poor, with improved mechanisms for ensuring their participation in technology development activities;

2 More cost-effective methods – drawn from the wide array of rapid and participatory approaches to rural appraisal now available[8] – of diagnosing the agro-ecological and socio-economic contexts in which small farm production takes place, and of monitoring and evaluating the performance of technologies and changes in management practice;

3 Enhanced awareness of the implications (especially for income distribution) of individual technologies; awareness of their input and processing/ marketing requirements and the need to assess the likelihood that these will be met;

4 More effective adaptation of technologies developed on-station to the conditions of the resource-poor farmer;

5 Enhanced concern among researchers to link with extensionists, primarily in order to obtain feedback on the usefulness of particular technologies in addressing specific issues, and to derive agenda for new research as further opportunities or constraints are identified.

In these relationships, NGOs might be expected to benefit from access to specialist skills and facilities that an 'opening-up' of government services to their needs would bring. Improved access to ideas generated in the wider scientific community and closer awareness of work being conducted elsewhere among NGOs were also perceived as potential benefits to individual NGOs of closer links with public sector research services. Overall, NGOs and ultimately their clientele of the rural poor, were expected to benefit from an increased volume of better-focused research conducted by government as a result of their interaction with NGOs.

In a similar vein, early outcomes of the ISNAR study of small country research systems (Gilbert and Matlon 1992; Gilbert pers. comm.) may be relevant to the question of NGO–government interaction around task specialization – although the context of the small country suggests that the tasks involved would be different. That study suggests that in small countries a brokerage function might be a more appropriate role for research institutes. This role might involve research institutes in the identification of local needs (of farmers and local support organizations such as NGOs), which they would then meet less by conducting their own trials (as is the conventional mandate of NARS), and more by searching elsewhere for innovations made by farmers, the private commercial sector, NGOs and special projects of various kinds – innovations which, where necessary, would be tested through an informal network comprising a similar multiplicity of agencies and individuals. Searches would also extend to the innovations and ideas generated in the research services and elsewhere within neighbouring countries, and at relevant international research centres. Formal testing of technologies acquired in this way would only be conducted where specific (and potentially serious) problems unlikely to be addressed through testing in the informal network would be expected to arise. In this new mode of operating, NARS would build on the advantage of their position in international and national information networks, and NGOs on their advantage of proximity to client needs.

Again such a change could only come with changes in the NARS involved: changes such as a substantial shift in the self-image of research staff, so that they see themselves as facilitators rather than elite researchers. Either way, these preliminary ideas point again to a division of tasks: the NGOs would give NARS information on their own stock of innovations, as well as requests for forms of technical support, and the NARS would bear the search costs implied by linking these two decentralized structures: one of demands for technical support, and one of potential supplies of technological solutions.

Proposals such as these for allocating tasks among institutions are now well beyond simple hypotheses. Across Latin America, Africa and (less so) Asia, a number of changes in the structure of the public sector, and the NARS in particular, have sought to promote an allocation of tasks among different institutions as part of a more general reduction of the levels of central government activity and funding in agricultural research and extension.

A number and range of changes have already been implemented within the context of this theme.[9] In Bolivia for instance, the NARS has withdrawn from extension, in the expectation that NGOs (and other private and local governmental institutions) will continue providing, and financing this service. The model proposed is that the NARS will provide technical assistance to NGOs as 'intermediate users' of NARS output, which NGOs use in their work with the final clients of research (the farmers), and report back on to the NARS (Veléz and Thiele 1991; Bojanic 1991; Bebbington and Thiele 1993). In Chile a somewhat similar model is in place, although there the intermediate users are paid by government under contracts to give technical assistance (Aguirre and Namdar-Irani 1992; Sotomayor 1991; Bebbington and Thiele 1993).

In some cases, alongside this public sector retrenchment has come a decentralization of tasks (and corollary financial responsibilities) within the public sector. In the Philippines, Colombia and Nepal (see Box 2.3), as just three examples, such a decentralization process is under way. In Colombia, extension responsibilities are to be passed on to municipal authorities who will contract or co-ordinate with agents (including NGOs) to give technical assistance (Wilson 1991). The role of the NARS will be to train and pass technology to these agents. In the Philippines, research responsibilities are allocated to area-based consortia covering the entire country. Each comprises at least one university, together with government research stations and, in some cases, NGOs. Problems which cannot adequately be tackled by the consortia can be referred to central government institutes specializing in certain commodities (Villegas pers. comm)

In yet other cases, state withdrawal is less, or not apparent. None the less, moves have been made, or are being considered, to involve NGOs as a means of improving feedback to, and exerting pressure on, NARS. Examples include the work of NGOs in collaboration with the Ubon Farming Systems Research Institute in N.E. Thailand (Sollows *et al.* 1991).

INSTITUTIONS, HISTORIES AND ACTORS: NGOs BETWEEN THEORY AND RELEVANCE

This functional approach to agricultural development institutions can thus take us some way to understanding what it is that governments and NGOs do well, and how tasks could be 'efficiently' allocated among them according to comparative advantage. However, a further theme that requires attention is how relationships between and within organizations actually operate. An understanding of this is important both for analysing the relationships between NGOs and governments, and then structuring those relationships if functions in the ATS were to be shared.

Perhaps more importantly, these functional prescriptions beg several questions: why is it that donors and governments are only now identifying these functional complementarities; why, if these organizations have these strengths and weaknesses, have they not joined forces in the past; and why is it that in many cases there has been resistance from some of the organizations involved to proposals for such a subdivision of tasks? For indeed, in several cases where governments and donors have asked NGOs to participate alongside government institutions in agricultural development programmes, NGOs have been critical and hesitant.

To answer these sorts of question, it is helpful to view the institutions involved not simply as organizations that execute functions, but as organizations of human agents which themselves have identities, objectives, and above all, social histories. Similarly, one has to address how institutions (and individuals within them) form views of the world around them, and of the role that they and other organizations do and should play in that world. Furthermore, the problem of power necessarily must enter the analysis. For if the issue of institutional reorganization is on the agenda, the question arises as to who has the power to enforce it, and who to resist it. As the ISNAR studies recognized, pressure from interest groups external to the NARS (such as donors, farmer organizations and policy makers, for instance) influence the orientation of research and extension work done in the public sector, and the nature and quality of the linkages between governmental research and extension institutions. Similarly these pressures influence the linkages between GOs (government organizations) and NGOs, and the ATD work done within NGOs.

These sorts of question necessarily require us to analyse NGOs and NARS in the context of the relationships that have structured policy and political contexts, and the distribution of power and resources in societies. This sort of comment, however, raises alarm signals among those who are interested in doing research that decision makers can use.[10] They fear that these types of sociological question rarely generate answers that are relevant to development agencies (Edwards 1989). Concepts of class, power, state, civil society and fiscal crisis by themselves give decision makers no obvious peg on which to hang their actions.

In some regards this sociological concern may have influenced the ISNAR studies. Some of the essays in that work are quite explicit in dealing with issues of power, pressure groups, institutional culture, and the relationship between human actions and the reproduction of research and extension institutions (Kaimowitz et al. 1990). They also deal with political economic influences on research and extension organizations. Martínez Norguiera (1990) in particular locates public sector research and extension activities within general considerations of the changing role, and structure, of the state in macroeconomic development. In Latin America, he argues, the increasing state intervention from 1930 through to the 1970s required increased complexity in the structure of government, to an extent that has now become difficult to manage. Indeed, he suggests that the difficulty has become such that in the current fiscal crisis the state has attempted to 'reduce internal complexity and divest itself of some of the responsibility for agricultural technology by handing over research and extension functions to the private sector' (1990: 97).

This contextualization helps the reader understand the political and economic dynamics of institutional change, and the general structure of management problems to which this gives rise. Indeed several of the ISNAR documents suggest the relevance of political economy and socio-political history for a proper understanding of inter-institutional and state–society relationships. None the less, ISNAR's primary emphasis remained oriented toward those factors that research managers could do something about.

It is a moot point as to how far analyses that elaborate structural and political economic themes are relevant to decision makers such as research managers. Such research may not give them entry points, but it may say a great deal about the environment in which they are making those decisions, and so say much about the 'room for manoeuvre' open to them (Booth 1992). By pointing to constraint as well as choice it can orient decision makers to those actions most likely to reward the time and resources invested in them, and most likely to meet their goals.

We would therefore suggest that analysis of this wider context is relevant research. Furthermore, we maintain that it becomes that much more relevant once a discussion of research and extension linkages goes beyond linkages

43

between different public sector institutions and includes relationships between them and NGOs. To appreciate why this is so, it is helpful to consider the social and political economic contexts in which NGOs have emerged.

Political history, institutional origins, and actors' strategies in the NGO–state relationship

In Chapter 1 we drew attention to the limitations of the label NGO. Even in the circumscribed way in which we are using the term (to refer to Carroll's GSOs and MSOs: Carroll 1992) there is still much diversity in the nature of these organizations. There are several axes to this diversity, among which are sectoral specialization, geographical cover, size, organizational structure and ideological orientation. Another very significant axis is that of the social origins of the NGO, and the motivations behind its original creation. These origins and motivations have a great influence on the identity of the NGO, and its attitude to the State.

In his own analysis of grassroots development organizations, Alfred Hirschmann (1984) has drawn attention to the ways in which individuals with a commitment to progressive social change tend to outlive organizations with the same commitment. Even if the organization disappears, or abandons its progressive orientation, the individual will find (or create) a new organizational form in which to pursue their personal political and ideological commitment. This effect he calls 'the conservation of social energy' (Hirschman 1984; cf. Ritchey-Vance, 1991). This leaves hanging the question as to how that energy, or commitment, was generated in the first place, but does help us understand the origin of many NGOs, or the forces that have brought individuals to choose to work in progressive NGOs.[11]

Many individuals currently working in NGOs began their professional life elsewhere. Indeed many, perhaps particularly in the case of Latin America, had previously worked in the public sector. Some of these moved into the state during periods in which it was pursuing progressive social development programmes such as land reform. In other cases they joined mainstream programmes. When these programmes were wound down, or when the individuals became increasingly disenchanted and frustrated with the limits of the public sector, many of them chose to leave the state (Carroll et al. 1991; Ritchey-Vance 1991). In other cases, when liberal governments were replaced by authoritarian regimes, they were forced to leave (Loveman 1991). A significant number of these formed, or moved to NGOs as a new institutional vehicle for pursuing their vision of social change (Carroll et al. 1991; Lehmann 1990; Loveman 1991). Work in NGOs also gave a source of income for some who, for political reasons, would otherwise have been unable to find work.[12]

The corollary of this situation, though it is far less frequent, is that some individuals can form right of centre NGOs when they find government policy too liberal or socialist, and create an NGO as a means of promoting and

44

pursuing a more market oriented form of development. Libertarian research NGOs, such as the Peruvian Institute for Liberty and Democracy, advocating the free market and the withdrawal of the public sector from diverse forms of intervention, are one such example (de Soto 1987).

More recently, in the context of public sector reform and neo-liberal economic policy, a different form of professional displacement has occurred that has been equally influential in the NGO sector. These adjustment related policy measures have not spared the professional middle classes (especially civil servants), who have seen their wages decline rapidly, and in some cases their jobs disappear. Some of this economically displaced middle class appears to have moved into, or created, NGOs in search of new higher paying jobs. This strategy has been aided by the increased availability of donor funding, which has facilitated the creation of new NGOs. In some measure, this has been an important element of the recent explosion of the NGO sector.

Of course there are many other ways in which NGOs can emerge, or in which individuals decide as part of their personal and professional evolution to join NGOs. For instance, individuals have moved into NGOs after having worked as researchers in university groups starved of funds, as student movement activists looking for means of pursuing more practical left wing politics, or as activists in political parties and unions.

The general point is that NGOs can emerge in different political economic contexts, and individuals can join or form them with different motivations. This leads to different attitudes to the state. Those NGOs that emerged in authoritarian contexts carry with them the experience of a state that can become repressive and violent – at times, indeed, violent against them and the rural poor. Many therefore take with them the experience of a negligent state that is not always to be trusted. In other cases, those who joined NGOs after becoming disenchanted with the bureaucratic and inflexible nature of the public sector carry with them a distrust of an inefficient state. Unsurprisingly, this sort of individual and NGO is much more circumspect about subsequent contact with the state.

One can be sympathetic to the distrust these individuals feel towards the state, and any contact with it. On the other hand, it is also the case that many of these individuals went into the NGO sector in order to develop work methods and development options that might one day be applied more widely (or scaled up) in a non-authoritarian and democratic state (Sotomayor 1991). Their work thus presupposed the idea of trying to change the state, and trying to influence and change public policy. For this reason, however much history makes a distrust and rejection of contact with the state understandable, this attitude cannot be accepted uncritically. In some cases it can be inconsistent with the motivation behind the NGO's creation. This is particularly so when the NGO is now faced with a state that has become more electorally accountable and decentralized, and less repressive and authoritarian than the state under which the NGO first emerged.

One can expect that this attitude of distrust and distance will differ significantly from the attitude of those individuals whose initial motivations for joining NGOs were pragmatic and economic rather than political. Such individuals are less likely to colour any decisions regarding contacts with the public sector with such distrust. Overall, their assessment of any such collaborations is likely to be less complicated (and mainly grounded on economic criteria) than that of those whose assessment will also be influenced by political criteria.

We do no more than point to the relevance of these individual and institutional histories to NGOs' policy with regards to contacts with the state. The actual form and impact of this history on a particular NGO's relationships to the state will depend on the particular context. Furthermore, this historical influence will be one of many. For instance, NGOs links to the state are not only institutional – they are also mediated by the social networks of individuals, which can often cut across the state–NGO boundary, particularly for those individuals who have moved from the state to NGO, or vice versa. The quality of these individual contacts may have more influence on any decision regarding inter-institutional co-ordination than generalized institutional relationships and images (see below).

Socio-political contexts of NGO–state relationships

Of course, the historical is not the only socio-political context influencing the relationship between NGOs and the state. However, how NGOs respond to contemporary contexts is contingent on the identity and experience that have been accumulated over time.

It is a truism to say that contemporary socio-political contexts of this NGO–state relationship are diverse and country specific. Furthermore, even if there is some pattern in the interaction between context and relationship, as we have suggested in the regional volumes (Bebbington and Thiele 1993; Farrington and Lewis 1993; Wellard and Copestake 1993), in the last instance this interaction takes a form that is case-specific.

While any generalization about the socio-political context of this inter-institutional relationship is therefore hazardous, let us at least point to several factors that influence NGOs' and the state's attitudes to each other, and thus that influence the potential for a closer relationship.

NGO attitudes to working with the state

We have suggested that NGOs formed under political repression develop an identity and culture that make it difficult for the NGO to trust and work with the state. While many suggest that the transition to democracy ought to lead to a change in this attitude, experience suggests that such changes are not straightforward nor automatic. There are a variety of reasons for this.

The more obvious obstacles stem from bureaucratic populism as a government strategy (see below) – elected governments are quite likely to perceive NGOs as a threat or competitor. Similarly, as experience shows recent returns to democracy have often been fragile, and at times more formal than real (de Janvry *et al.* 1989). During the course of this study, attempted military coups in Venezuela, and a successful civilian coup in Peru made this fragility quite apparent. Moves to democracy in several of the African case study countries were similarly very faltering.

A poignant example of the implications of this instability for NGOs and donor strategies to involve NGO comes from Togo. In its 'Togo Grassroots Initiative', the World Bank in 1989 agreed $5 million of credit to the government, £3 million of which was to be passed to NGOs for a range of NGO-managed community development activities. The World Bank (1990a: 12) praised 'The dynamic and diverse NGO sector in Togo and the government's recognition of the contribution that NGOs make to the country's development . . .'. The rise of political violence, and severe instability in 1992, however, call into question the possibility of any future relationship between the state and NGOs who are even mildly critical of its policies.

Just as the transition to electoral democracy does not solve all political problems, nor does it solve bureaucratic inefficiencies in the public sector. Consequently, some NGOs continue to be wary of co-ordinating with government. The fear is that in doing so the effectiveness of their own work will be reduced, and thus their relationships with the rural people with whom they work, upset.

In addition to the often well-founded reasons for caution on the part of NGOs, a more subtle but perhaps more profound obstacle to their engagement in closer relationships with government stems from the difficulty of redefining institutional identity. For those NGOs who have long understood themselves and their role as a form of political opposition and criticism, to begin collaborating with government in joint programmes, to begin implementing elements of public programmes, or to begin dividing tasks in the ATS all imply a quite profound change in identity. For some NGOs the change is impossible, and for others it can cause great tensions in the organization.

This challenge to institutional identity and cohesion is particularly acute in a country like Chile where NGOs have experienced, and indeed have played a key role in promoting an abrupt shift from authoritarian military rule to a reasonably consolidated and efficient electoral democracy. The new government's policy aimed to increase public expenditure on social development activities, but to contain any increase in government implementation of those programmes by contracting them out to the private commercial, and (largely) to the NGO sector.

In the agricultural sector, the National Institute for Agricultural Development (INDAP), responsible for technical assistance to farmers, contracts out extension work, and many NGOs have won these contracts. This however has caused problems for some, which can be illustrated by contrasting the experiences of two of Chile's prominent NGOs, AGRARIA and GIA (Grupo de Investigaciónes Agrarias). AGRARIA decided to accept a large number of INDAP contracts, perceiving this as an opportunity to influence the use of government funds for rural development and poverty alleviation. None the less, in doing so, it has had to take on many new staff, has almost doubled in staff numbers and has had to adapt those parts of it working on INDAP contracts to the contractual requirements of INDAP, some of which require a field methodology that is at odds with the traditional methods of AGRARIA. These factors have introduced imbalances in the NGO. The new staff are not always as ideologically committed, nor identified with the institution's own political trajectory as the older staff. Likewise, the organization has grown so large (around 120 staff) that it is now impossible for all members of AGRARIA to know each other and identify with its other field programmes. Furthermore, the institution runs the risk of internal division between the part that acts as AGRARIA has always done, and those parts that even within AGRARIA are referred to as 'AGRARIA-INDAP' (Aguirre and Namdar-Irani 1991; pers. comm.).

While these tensions have also occurred in GIA, the stress there has been somewhat less, as it has perceived such contracts as research rather than implementation exercises. That is, GIA implements INDAP contracts in order to study how they function, and on that basis make recommendations on how they could be changed and improved. Reviewing these experiences, Sotomayor (1991) (writing from GIA's perspective) suggests that GIA's response is the more viable for an NGO. In the long term, NGOs are more likely to retain an institutional identity that in some sense resembles their past, if they pursue a link with government in which they develop innovations for presentation to government, rather than one in which they implement public programmes. The more they commit themselves to the latter the more they may lose their identity and motivating vision. They will become increasingly the implementers of somebody else's programmes (Sotomayor 1991).

In short, Sotomayor (1991) suggests that under democracies such as that in Chile, where government takes on board lessons from NGOs and begins to approach them for advice and assistance, the most viable NGO identity for those NGOs that were formed under more difficult political conditions, and continue to identify with the need for redistributive social development, is as 'innovator' and 'constructive critic' rather than as 'opposer' or 'implementer'. Of course, for the 'opportunistic NGOs' created much more recently, and whose reason for existing is largely to implement programmes rather than to affect any social change, then the identity as implementer might be much more acceptable.

In addition to the threat to their *identity*, closer contracted relationships between NGOs and government present NGOs with a number of other challenges. Their *financial security* is undermined, as the more that NGOs accept government funded contracts, the more they are vulnerable to the winds of change in the politics of government. Their *autonomy* is challenged: by engaging in co-ordinated programmes, NGOs surrender a certain degree of autonomy over their own actions and the external factors that might affect them. Finally, their *cohesion* is weakened, as new tensions emerge within the NGO.

For all these reasons (and more), 'whether and under what circumstances a particular local organization should relate to government or avoid government is a major strategic decision' (Esman and Uphoff 1984: 267). And yet, as we have also noted, much of the reasoning behind the work of NGOs concerned with social change is that the institutional, policy and technological inno-vations they develop should one day be implemented on a wider scale, which necessarily means through a public sector structure. Thus Esman and Uphoff go on to conclude that despite the risks of contact with government, 'general prescriptions for absolute autonomy, for implacable confrontation with the State . . . are not suited to trends that are evident in rural life' (ibid, p.267).

The state's attitudes to work with NGOs

In the simplest sense, one might expect the state's attitude to NGOs to be the mirror image of the NGOs' opinion of the state. It might therefore be expected that those NGOs which remain apolitical and market oriented will attract little adverse attention from the state, whereas NGOs that are politically critical of the state will face greater likelihood of repression (Fowler 1990; Lehmann 1990). In such circumstances, no amount of functional complementarity will usher in an easy working relationship between the two.

There is much evidence that these political factors do indeed have a great influence on the state's attitude to NGOs. In Africa, Bratton (1989: 572–6; quoted in Fowler 1988: 57) concludes: 'the amount of space allowed to NGOs in any given country is determined first and foremost by political consider-ations, rather than by any calculation of the contribution of NGOs to economic and social development'.

In cases, then, where the demand from NGOs and grassroots organizations for greater participation in setting and implementing the development agenda is growing, but where the state is resisting the introduction of democratic reforms (as in some African, but also in some Asian countries), then the state looks unfavourably on NGOs that show any hint of political opposition. On the other hand, there are cases where even authoritarian states benefit from the effective subsidy they gain from those NGOs implementing social and development services. This subsidy may lead the state to tolerate (within reason) the activities even of those NGOs politically opposed to it. One may expect this situation to occur more in stronger states which are more confident in their ability to withstand NGO criticism. This may, for instance, explain why Pinochet's Chile tolerated (within limits) this NGO social work (Lehmann 1990: 81).[13]

More significant, however, is the fact that there is no necessary reason why democratic states should look favourably on NGOs. For instance, even if the NGO is not, and has never been, overtly critical of the state, its actions in social development imply criticism of the state's shortcomings, a criticism not lost on the rural poor (Bebbington 1992; Fowler 1990). For Africa, Fowler (1990: 63) goes as far as to suggest that 'this situation is probably one of the most significant sources of political tension between existing regimes and the voluntary sector as the latter expands'. Such tensions are aggravated yet further when the NGO sector is dominated by opposition political parties.

A related source of tension, of course, can arise from the state's resentment over the volume of resources it may see as 'diverted' to NGOs by donors. Indeed a number of governments have tried to capture or restrict this flow of finance to NGOs.

Thus the state may choose to pester and repress, control or co-opt even apolitical NGOs. On the other hand, other forces may be pressing the state to engage with NGOs. Donor pressure on states to pass rural and agricultural service provision to the private sector, and specifically to NGOs (Williams 1990; World Bank 1990a), is one such force encouraging the state to recognize and work with NGOs it might sooner weaken. One packet of 'solutions' to the implied problems this presents to the state might include some combination of restrictive legislation (as in the NGO registers proposed in Kenya in 1987, and in Bolivia in 1988), administrative co-optation and political appropriation (Fowler 1990: 64–9). More specifically, attempt might be made to respond to this pressure by creating a contractual relationship between the state and NGOs, which might open up to the state a number of opportunities:

1 To use such contracts to reduce the flexibility of NGOs, and make them increasingly oriented to the state's demands as contractor rather than to the demands of the rural poor;
2 To use contracts to co-opt NGOs politically;
3 To create government organized NGOs (GONGOs) and then award them contracts;
4 To package the field presentation of programmes as public sector programmes that are merely implemented by NGOs. In this way the state may gain the legitimacy stemming from service provision, and yet divert any criticisms to the NGO.

It is for reasons such as these that Montgomery (1988) refers to NGO–GO co-operation as a strategy of 'bureaucratic populism' in which the state aims to co-opt NGOs in such a way as to counteract the erosion of public trust in government and help it (the government) achieve its policy goals.

Thus, even if 'the image of the monolithic regime conspiring with rural elites in the systematic exploitation and repression of the rural poor hardly describes the orientation of most governments' (Esman and Uphoff 1984: 281), nor does democratization remove barriers to NGO–state collaboration at the stroke of a pen.

None the less, we should heed Esman and Uphoff's caution to guard against this 'image of the monolithic regime'. Many developing country states are too weak to be monolithic, and as adjustment-led public sector reform measures proceed apace as we write, these states are becoming weaker still. Furthermore, it should be noted that in many instances, there can be a very strong case for greater state influence over NGO activities – particularly in circumstances where NGOs act in an uncontrolled and uncoordinated fashion.

State registration of NGOs

Both the good motives for state control and the bad lie behind requirements for NGO registration. Consequently, the way in which such registration is managed by the state is often an important influence on the quality of NGO–state relationships, and possibility of fruitful interaction.[14]

The justification given by governments for their efforts to regulate NGOs fall into two broad categories: issues related to the need for financial control, and the need for co-ordination of development activities.

The need for financial control

At its simplest level, such control requires NGOs to keep accounts of income and expenditure. However, once detailed accounts are submitted, they allow government to check individual items of expenditure (the level of salaries and allowances paid to expatriate staff is a recurrent object of scrutiny) and to check sources of income – the amounts and origins of foreign exchange receipts being of interest to most governments. Once detailed information of this kind is available, it requires only a short step for government to try to *control* income and expenditure.

It would be naive to suppose that financial and development motives were the only ones underlying governments' registration efforts. As Box 2.1 indicates, in Kenya they have been accompanied by arbitrary action (and threats of action) by government against NGOs whose commitment to empowerment and democratization they regard as subversive (Wellard and Copestake 1993; Fowler 1990).

In India (Box 2.2) the Foreign Contributions Regulation Act has given government the legal basis for far-reaching investigations of NGOs whose views and actions have challenged the government's own (Robinson *et al.* 1993). In Nepal, the Social Service National Co-ordination Council has similarly been used to screen any NGO activities that might be construed as political, and, in some cases, to promote NGOs that offer employment opportunities to government's political allies (Box 2.3).

Co-ordination of development activities

A further broad reason for registration is the rising concern among governments that, as levels of NGO activity increase, they should be at least broadly consistent with government development measures. Even within the framework of this requirement, liberal governments will be sensitive to NGOs' potential for developing alternative strategies, approaches and methods which may offer wider lessons; the less liberal will require that NGO activities should be fully co-ordinated with those of government. A number of governments, often as part of a strategy of administrative decentralization, have begun both to allow NGOs an increasing voice in setting local development priorities and, at the same time, to require NGOs to seek the approval of local administrations for their proposed activities. This is clearly an element of recent reforms of local administration in Nepal (Box 2.3).

Box 2.1 Government mechanisms for NGO co-ordination in Kenya

Following the attempted coup in 1982, the Kenya government's previously open and accommodating attitude towards NGOs gradually hardened to the extent that in 1989, for instance, the national women's organization (Maendeleo ya Wanawake) was affiliated to the national President's party by decree. Similar threats have been made against the National Council of Churches, and the more radically empowering NGOs, particularly those affiliated to the Catholic Church, have come under close scrutiny from the internal security services. In 1991, after only cursory consultation with NGOs themselves, a new framework was established for compulsory registration of NGOs with a statutory Board accountable to the office of the President. Efforts to ensure compatibility of NGOs' efforts with local development plans include the requirement that NGOs should submit proposals for expatriate recruitment and for imports to the relevant District Co-ordinating Committee for approval. Additionally, NGOs are required to co-ordinate their activities at district level with those of government under the 'District Focus for Rural Development' strategy established in 1985. However, the co-ordinating performance of this strategy has been highly uneven.

Actions such as these have led to views (*Kenya Daily Nation*, 26 June 1992) that the Co-ordination Board created under the NGO Act will continue to be treated with suspicion by NGOs.

Source: Wellard and Copestake 1993

Despite the dangers of politically-motivated control inherent in requirements of NGO registration (Boxes 2.1–2.3), it is clear that there remain valid developmental grounds for some degree of co-ordination between government and NGO efforts. This is especially the case in those areas where large

---

Box 2.2 Government mechanisms for NGO co-ordination in India

NGOs in India are required to register under the Societies Registration Act of 1860. The Foreign Contributions Regulation Act of 1976 (extended in 1984) governs the receipt of foreign funds, which contribute around 90 per cent of NGOs' income. This legislation does not prevent large volumes of foreign funds from entering the country illegally, but has served as a pretext for some government investigations of NGOs allied with the political opposition, as it did under the Kudal commission of enquiry in 1982 (Robinson 1991). Party political considerations therefore influence government's relations with NGOs to some degree, but its overall attitude is positive. The Council for the Advancement of People's Action and Rural Technology (CAPART), for instance, was established in 1986 with the twin aims of promoting NGO involvement in rural development and promoting technological innovations through NGOs. In practice, it serves as a channel for distributing some of the India Government funding (totalling US$170 million/year on average in 1985–90) to NGOs, including part of the funds received through official bilateral assistance.

Source: Robinson 1991

---

numbers of NGOs have initiated development activities which overlap and conflict both among themselves and with those of government.[15]

There can be little doubt that NGOs would find it easier to defend their interests against undue interference from government, and that both they and government would be better able to capitalize on sensitively-managed interaction between the two sides if NGOs themselves had a better record of working with one another. However, their highly individualist philosophies and modes of operation has made inter-NGO co-operation more often the exception rather than the rule.[16]

Policy contexts of NGO–state interaction in the case study countries: a rash synthesis

These are just some of the many contextual issues that need to be borne in mind as we consider the possibility, nature and implications of NGO–state interaction. The diversity of issues warns us once again to remember that each country will have its own particular context with its own particular implications for NGO–state interaction.

However, if we look at the countries in which research was conducted, some (rash) generalizations might be made about the overall context of the policy agenda – poverty orientation, the capacity to implement poverty-focused policies and overall government attitudes towards NGOs – which government sets for NGOs.[17] While deriving largely from subjective assessment of governments' policies towards the rural poor, and their capacities to implement them, the three broad types of context etched in the following

---

Box 2.3 Government mechanisms for NGO co-ordination in Nepal

Voluntary action at the local level has a long history in Nepal, and continues to be an important means of providing facilities (bridges, trails, canals, schools, temples) where communities remain isolated. In the 1950s–1970s, government was hostile to the establishment of new, larger NGOs unless they could serve to draw in foreign funds *and* provide employment opportunities for the elites who supported the political system. Compulsory registration of NGOs with the newly-established Social Service National Coordination Council (SSNCC) under the patronage of the Queen was introduced in 1975, with strengthened procedures in 1978, but a number of NGOs are still able to operate without registering. The new constitution of 1990 removed some of the SSNCC's powers of patronage and control, but a recently drafted government directive envisages a new range of financial controls over NGOs, and visa restrictions for expatriate NGO staff, to be implemented through a re-strengthened SSNCC. Recently approved local government bills envisage NGO representation on Advisory Committees to be set up by each Municipality and Village Development Committee. NGOs will also participate in the design and implementation of local development activities, with funds both from increased local government budgets and from their own resources. Local government bodies are given the powers to ensure that the activities of NGOs are consistent with the agreed local development framework, to insist on co-ordination of activities among NGOs and to audit their accounts. Although the extent to which this legislation will be implemented remains to be seen, it is clear that closer co-ordination of NGOs' activities with government development plans is anticipated, making available to NGOs some government resources for these purposes, but also envisaging that substantial parts of NGOs' own funds will be allocated to collaborative activities in this way.

Source: Shrestha and Farrington 1993

---

paragraphs help to indicate how far different governments are open to the prospect of working with NGOs, how far the strategies and policies of NGOs and governments towards the rural poor are likely to converge, what room for manoeuvre NGOs are likely to have in their relations with government, and what broad forms of interaction are likely to emerge in specific circumstances.

Medium or low government policy orientation towards the rural poor, but strong government presence in rural areas; moderate to negative attitude towards NGOs (e.g. Indonesia, Thailand, Kenya).

The low policy commitment towards the rural poor in this group is explained by the high degree of control over the policy agenda exercised by other more powerful groups (agribusiness in Thailand; cronyism and tribalism in Kenya; the military in Indonesia). Governments see themselves in firm control over developments in rural areas and while, in the case of Thailand, willing to permit the widespread establishment of farmers' organizations in such functional roles as input supply and marketing (Garforth and Munro 1990), regard non-membership NGOs, especially those having strong foreign links, with some suspicion. While the scope for links between the two sides is limited by these conditions, neither NGOs nor governments are monolithic and, as the case studies reported in companion volumes indicate, adequate levels of openness and diplomacy are sometimes achieved to permit fruitful interaction.

Medium government policy orientation towards the rural poor, but weak government presence in rural areas; moderate/positive orientation towards NGOs (e.g. Ghana, The Gambia, Senegal; most of the Latin America case study countries).

Weak government presence in rural areas coupled with a broadly positive attitude towards NGOs is likely to give NGOs considerable latitude in developing their own programmes and in linking with government services where there is clear perceived advantage. In these conditions, prevalent patterns of interaction in ATD might be expected to include NGO adaptive trials on and dissemination of technologies developed by government. The danger, of course, is that government frustration over its inability fully to implement policies towards the rural poor may lead it (and donors) to view NGOs as substitutes for government in service delivery and, as discussed below, this would be seen by the more innovative NGOs as excessively restrictive and would leave their talents unexploited. The Philippines may be regarded as a close outlier of this category. It differs insofar as major indicators such as its poor record on land reform suggest a weak policy orientation towards the rural poor. However, the provisions for inclusion of NGOs in development processes enshrined in the Aquino constitution, and the moves to open up the administration to interaction with NGOs (see Miclat-Teves and Lewis 1993; Fernandez and del Rosario 1993) have provided a particularly favourable climate for NGOs. Again, government is not monolithic, and a number of illuminating NGO–GO interactions have taken place, as the case studies indicate.

Medium government policy orientation towards the rural poor, but weak implementation capacity; moderate/negative environment for NGOs (e.g. Bangladesh, Nepal).

The high incidence of rural poverty in these countries, together with weak government capacity to implement development initiatives, undoubtedly provides fertile ground for NGO activity. However, registration requirements imposed in one of these countries (Nepal – see Box 2.3) have been found excessively restrictive by many NGOs and, in both, NGOs rarely find willingness and/or capacity among GOs to offer anything useful in a partnership context. Not surprisingly, the case studies show only limited positive links between the two sides in these situations.

These are but broad generalizations, offered in much the same way as Tandon 1987 (referred to in Clark 1991) delineates three broad regime types that offer different environments to NGOs: the dictatorship, the single party state, and the liberal democracy. Whatever categorization one uses to allow at least a broad structure in which to organize one's thinking about NGO–state relationships, the categorization must be used cautiously and flexibly: for instance, countries move from one category to another as national political circumstances change. And so with due caution, and no more ado, we therefore move back to a perspective on NGOs and states that reminds us once again of diversity: an actor-oriented perspective.

An actor-oriented view on NGOs and NARS

In several parts of the discussion we have drawn attention to the relationship between institutional dynamics, and the strategies of individual actors. The strategies that institutions pursue, and the ways in which they position themselves in civil society and in relationship to the state are an effect (albeit not linear) of a complex of individuals' strategies.

Much of the remaining text will be pitched at the level of organizations' actions and interactions, rather than those of individuals. Given the focus, scope and length of the book this degree of aggregation is perhaps inevitable. None the less, in various cases, particularly when we discuss NGO–NARS linkages, we will draw attention to the ways in which contacts between individuals arising out of shared backgrounds, or informal contacts at field or professional levels, can lead to collaborations between organizations who would have found it difficult to work together had the interaction been initiated at a formal institutional level. Beyond the question of interpersonal contacts, socialization into particular structured practices (of class or ethnicity, say) also facilitates communication with, and implicit understanding of other individuals socialized into similar sets of rules and norms of social behaviour. Agents' practical ability to access other agents is thus differentiated. A typical NGO worker who is urban educated and has middle class family origins[18] may then have easier access to similarly socialized individuals in the state than to the ethnically different peasant farmer to whose emancipation the NGO is ostensibly committed.

It is also the case that individuals have a room for manoeuvre within their organization that can lead them to draw it in certain directions. More generally, institutional reproduction is a net effect of individual actions, and so an understanding of an organization's actions, culture, and how these are changing, would ultimately need to be based on an analysis of the actions and relationships between individuals in that organization. Such an analysis would also illuminate how many of those actions are part of strategies pursued by actors which have only a limited relationship to the stated (mandated) objectives of the institution on whose resources they draw when they act (Long and van der Ploeg forthcoming; de Vries 1992).

Such institutional ethnographies have much to contribute to our understanding of what NGOs and NARS are and do, and how they interact. They would undoubtedly call into question some of the (and our) more cavalier generalizations about the motivations and objectives of these organizations. However, they would also draw attention to how change occurs in institutions, and thus to how decision makers and managers might be able to effect strategic change. In this way, while academic in one sense they would also be relevant research for decision makers.

While there are some ethnographic and actor based accounts of the work and dynamics of government institutions (e.g. de Vries 1992; Grindle 1986), few have been written about NGOs. Our comments are based on our knowledge and the case studies written by the NGOs who participated in this study. We draw attention to this area as one urgently requiring research in the future as NGOs become responsible for ever larger amounts of social development expenditure.

The problem of diversity

Entering the social history of NGOs in this way, and their particular relationships with and situation within state and civil society, raises the problem of diversity. Social and institutional histories will differ among the three regions, and among and within countries in each of the regions. Added to the similar diversity within the label of NGO, we are left with the very difficult question of how far generalizations can be made about NGOs, and their relationships to state and civil society (cf. Fowler 1990).

Of course, most policy and decision makers want generalizations that they can apply widely. The emphasis on diversity might, therefore, expose this approach to criticisms that it is 'irrelevant', or at least, not very useful. However, the implication of the actor-oriented view on these institutions is that this diversity is in part explicable as an effect of choices made by agents. These choices are not entirely free, but nor, so our analysis has gone, are they entirely determined. Decision makers in NGOs and government have room for manoeuvre (Bebbington and Farrington 1993). Thus the emphasis on diverse outcomes might be defended as relevant – it helps bring the actor back in (cf. Booth 1992).

Furthermore, we have suggested that actions and institutions need to be considered within, and as constitutive of, structural context – or, in other words, of pattern. Some general statements can be made about these patterns, and thus about the types of room for manoeuvre open to decision makers in particular structural contexts. Our discussion of the different origins of NGOs and the implications these have for NGO–state relationships point in this direction. So does our recognition that NGOs tend toward certain institutional strengths and weaknesses.

In this sense, the analysis in this book aims for relevance, points to patterns, but consistently warns (and, some readers may decide, annoyingly so), about diversity. Perhaps what our analysis is suggesting is that we can generalize about the sorts of questions that should be asked in particular situations, but not about the explanations of outcomes which necessarily remain context-specific. As opposed to applied studies that look for general models that can be applied in diverse situations, we are interested in general questions that can be asked of a particular situation. David Kaimowitz's work on research–extension linkages at ISNAR led him to a similar conclusion: 'Experience has shown, however, that it is impossible to come up with a set of general recommendations which would be appropriate in all circumstances. Solutions which work well in one context perform poorly in others' (Kaimowitz *et al.* 1990: 227).

However, we would suggest that there is one policy relevant conclusion of this argument that is generalizable. If local contexts are important in understanding the nature of inter-institutional relationships, then the closer decisions are made to those contexts the more likely they are to be appropriate. This, of course, is an argument for decentralizing research and extension institutions within countries – but it is also an argument for decentralizing many other aspects of development decision making.

CHAPTER SUMMARY

Many applied studies of ATD have focused on how to reorganize institutions and interrelationships among them and among researchers and extensionists in order to achieve a more efficient execution of the tasks that are required for effective ATD. This form of analysis has recently led to arguments that NGOs are peculiarly suited to accomplish downstream tasks of adaptive research and technology transfer, and to promote farmer participation and feedback to researchers. They go on to suggest that this NGO strength complements the special strengths of public sector institutions. They therefore recommend a division of tasks, or functions, among NGOs and government institutions.

The strength of such function-oriented approaches to institutional questions is that their results are more easily translated into actions that research managers can take. Their weakness, however, can lie in a failure to consider the historical and socio-political factors that similarly can influence inter-institutional relationships, and confound them no matter how strong a functional complementarity looks on paper.

Consequently, function-oriented analyses similarly need to be complemented by sociological analyses of relationships between and within institutions. We suggested that these analyses should pay attention to the importance of context, the reality of diversity, and the recognition that outcomes are an effect of the particular workings of human agents in structural contexts.

These analyses run a corollary risk of being 'too theoretical' and divorced from the operational world of decision makers. This need not be so, however. Such analyses should maintain a focus on tasks in the ATS, thus providing entry points to decision makers. But they should always encourage those decision makers to think beyond the micro-level, the task, and the immediate present, and think through the factors that will impinge on her or his room for manoeuvre and impinge on the outcome of these decisions.

The remaining empirical chapters are offered in this vein. Some pay greater attention to the functional strengths and weaknesses of NGOs; others stress the socio-political factors behind NGO actions in ATD. The image they present is the broad outline of the tendencies we believe we have observed in this research; the message they convey, however, is that generalizations – including ours – must be treated with caution.

NOTES

1 The ISNAR studies were not only couched in the terminology of functional and systems analysis, but they also paid attention to wider socio-political factors (what they called contextual factors: Kaimowitz *et al* 1989). It remains the case, though, that their own particular client orientation led to a focus on function and activities.

2 NARS is the acronym for National Agricultural Research Service. It was coined by the International Service for National Agricultural Research (ISNAR), one of the research centres within the Consultative Group on International Agricultural Research (CGIAR). In this book, it is used as convenient shorthand for public sector research and extension services.

3 Such single site locations arise, for instance, with research stations having a national or regional mandate for a particular export commodity such as the Cocoa Research Institute of Ghana – formerly the West Africa Cocoa Research Institute.

4 Farrington and Mathema (1991), for instance, quantify these for agricultural researchers in Nepal.

5 There is a corollary point to this. When per diems are available, workers often go on trips to supplement their wages; the effect is sometimes to draw them away from other work needing attention.

6 For Asia, for instance, this story is told in several of the papers presented at the conference on farming systems research held at the National Academy of Agricultural Research Management, Hyderabad, India, in October 1991.

7 A 'small country' has been defined as having a population of fewer than five million at the 1990 census, and meeting at least three of the following four criteria: the economically active agricultural population is 20 per cent or more of the total economically active population; per capita income is less than US$2,000 (1980 US constant dollars); AgGDP per capita for the economically active agricultural population is less than US$2,000; AgGDP is 20 per cent or more of GDP (Eyzaguirre 1991).

8 For example, see the *RRA Notes* published by the International Institute for Environment and Development.

9 Some of these changes and their implications for linkages between NGOs and NARS are discussed in more detail in Chapter 5.

10 Thus in ISNAR's case, there was clearly an effort to respond to this concern by generating information that could be useful to research and extension managers.

11 The question of the origins of individuals' social commitments, of their expertise in organizing and launching political strategies, and of their capacity to manipulate and negotiate with non-local institutions is beyond this book. However, we would make the following observation. Many explanations of the rise of NGOs and social

movements analyse them as reactions to structural changes – for instance, as reactions to authoritarianism, or to the failure and withdrawal of the state from service provision. While these were and continue to be very important factors in explaining these organizations, a full analysis would also need to address the social processes that brought individuals to the point where they wished, and were able, to organize in these ways. This would take the analysis, for instance, into questions of social network formation, impacts of education, and the development of popular political culture. Conversations with Roger Riddell helped clarify this point.

12 Daniel Rey, a co-founder of the Chilean NGO AGRARIA comments: 'The [Chilean] dictatorship gave rise to institutions that resolve – and here we must be honest – not only the needs that social groups have, but also the problems of professionals – who had no place to work; not only no place to work in the field we wished, but no place to work at all' (quoted in Loveman 1991: 10).

13 Lehmann (1990: 181) captures the irony of this situation: 'And as for Pinochet, he does not complain at all: have not the international agencies taken over his responsibility for the poor, and relieved him of the political pressure to change his policies. It is a most bizarre convergence of interests.'

14 The registration of a new NGO requires a minimum of formality in some countries (e.g. Philippines), but, at the opposite extreme (e.g. Thailand) it is a lengthy process requiring the approval of several government departments, including clearance of the principal office-holders by the criminal investigation department.

15 Examples include those in the Krishna delta of Andhra Pradesh, India, cited by Robinson (1991), the case of Siaya District in Kenya (Charles and Wellard 1993), and San Julian in Bolivia (Ayers 1992; Bebbington and Thiele 1993).

16 The few examples found within the case studies include that on indigenous seeds in Zimbabwe (Chaguma and Gumbo 1993). A further example is found in the Alternative Campesino (Small Farmer) Development Programme (PROCADE) which has a programme for inter-institutional collaboration among twelve NGOs working in five administrative Departments of Bolivia. PROCADE and the Bolivian NARS (IBTA) have attempted to work together – indeed, PROCADE represents NGOs on the Managing Council of IBTA – but collaboration has been impaired by jealousies on both sides, and by the lack of flexible mechanisms for inter-institutional co-ordination (Gonzalez 1991; see also Bebbington and Thiele 1993).

17 Bangladesh, India, Indonesia, Nepal, Philippines, Thailand, The Gambia, Ghana, Kenya, Senegal, Zambia, Zimbabwe, Bolivia, Chile, Colombia, Ecuador, Peru.

18 It became clear during the preparation of material for this study, that NGOs are generally administered by such people – markers include anything from surnames to travel labels on their briefcases. Fowler (1990: 73) also comments on this.

3

NGOs AND AGRICULTURAL CHANGE: TECHNOLOGIES, MANAGEMENT PRACTICES AND RESEARCH METHODS

INTRODUCTION

This chapter examines hypotheses arising from the previous two chapters regarding NGOs' role in agricultural change. It focuses on (1) their perspectives and actions in agricultural 'technology' – by which we mean both hardware for agricultural development (seeds, equipment, etc.), and also methods of managing resources, both on- and off-farm – and (2) their approaches to the development and dissemination of these technologies.

The chapter is organized as follows: first, we examine definitions of 'agriculture' and 'technology' and ask whether we might expect NGOs' definitions to be substantively different from those of government.

Second, we give a brief overview of the types of ATD work undertaken by case-study NGOs which allows us to assess the hypothesis that NGOs are best suited to the adaptive and transfer sub-systems of the Agricultural Technology System (ATS). Specific case studies presented in more detail illustrate the range of NGOs' perceptions of 'technology' and examine how far NGOs link their technological to their political and social agenda.

We will argue that there are two main tendencies in NGOs' approaches to technology, which we denote as *production-oriented* or *agroecological* (Figure 3.1). The former reflects a broadly sympathetic perspective on the role of new varieties and agrochemicals in small farm production; the latter reflects a preference for production relying on low or zero levels of fossil fuel-based inputs. A particular predisposition towards technology does not, however, have any necessary relationship to the NGOs' perspective on development. Thus, those working with 'modern' technologies may follow approaches that are very orthodox (high input), 'transferring' technical packages from one setting to another, but they may also pursue what we call grassroots-sensitive approaches, that is, drawing only on those technologies, dissemination methods and institutional forms which are appropriate to local conditions, and at times using Green Revolution technologies to very radical and progressive effect. Similarly, agroecological approaches may be either pragmatic responses to local opportunities and constraints, rather than reflections of an ideological

Technological focus   Wider context of technology focus

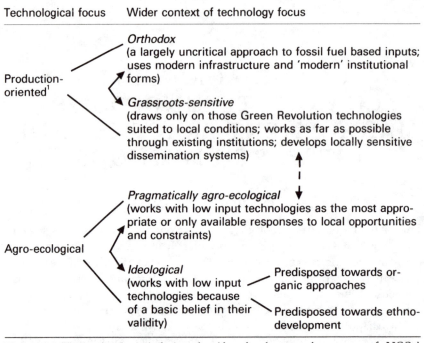

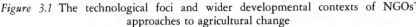

Figure 3.1 The technological foci and wider developmental contexts of NGOs' approaches to agricultural change

Note: 1 While in Chapter 1 we suggested that the modernization/modes of production dichotomy was less relevant in current debates, 'production-oriented' perspectives on technology are modernizing to the extent that they refer to the high input aspects of the Green Revolution.

commitment. Some, however, are grounded in ideological predispositions, among which two are especially important: a predisposition towards organic agriculture or environmental protection, and another towards the recovery and revalidation of the culture, social forms and technical practices associated with particular ethnic groups.

We analyse the case study material against this framework, and use it to draw out further issues of particular relevance to NGOs' experience. In particular we discuss (1) the extent to which NGOs draw on and seek to strengthen indigenous technical knowledge (ITK) in the process of designing innovations, and (2) contexts where they feel that their clients' interests lie on the spectrum between close interaction with a market system at the one extreme and subsistence production at the other. In the third section, we return to our earlier discussion of differences between NGO and government perceptions of agriculture and technology, and ask whether certain concerns are shared among NGOs independently of whether their focus is broadly 'production-oriented' or 'alternative'. We suggest that such concerns might include participatory methods and, as a prelude to closer examination of

NGOs' interactions with their clients in Chapter 4, we examine how far they link their work on technology to certain social concerns, such as the creation of income-generating opportunities for women and the landless, or the promotion of technologies capable of generating greater social cohesion among the rural poor.

Our discussion in the final part of the chapter falls into two broad areas: first, we emphasize that NGOs' philosophies and the approaches to ATD through which they are articulated are neither uniform nor static: they have evolved with field experience, with changes in the political, economic or institutional context, with increasing awareness of the wider changes in thinking on development and the environment, and through pressure from donors. Second, we raise questions about NGOs' costs and performance in relation to those of government in the various parts of the ATS in which they have chosen to work.

SOME DEFINITIONS

The definitions of both *agriculture* and of *technology* in this book are broad.[1] We define technology to embrace:

- Hardware (e.g. seeds, vaccines, machinery);
- Management practices and techniques (e.g. soil and water conservation practices, rotations, crop mixes, agroforestry);
- Increments in knowledge (whether traditional, modern or some combination of the two) that strengthen local capacity for experimentation, communication and general resource management.

While all three of these elements are of concern to both NGOs and governments, NGOs' concern with participation and empowerment suggests that the third – especially the development and sustainability of local knowledge systems – would be of particular interest to them. While improvements in each of the three increase production efficiency, the impacts of such improvements are unlikely to be neutral among different groups in the population. Thus, our interest in this chapter lies in both the production and the distributional impacts of the technical change promoted in the work of differing NGOs. The issue then arises as to how far NGOs' organizational characteristics also influence the nature of these production and distributional impacts. We touch on this theme in this chapter, but develop it in greater detail in Chapters 4 and 5 – where we also consider how far their structure and performance might be improved through changes internal to the NGO and in particular through links to GOs.

If, as was suggested by literature reviewed in Chapters 1 and 2, NGOs tend to take a holistic view of issues affecting their clients (in part because they are not constrained by mandates), then we might expect their conception of agriculture to go beyond the simple production of crops, or of crops, animals and trees, and to embrace the interactions among these different components of production systems. We might also expect them to be concerned with the interaction between on- and off-farm use of natural resources given the strong empirical evidence[2] that poorer households rely more heavily on access to and use of open-access or common property resources. Such reliance embraces a wide range of economic activity including, for instance, the collection and sale of minor forest products and wild foods, the collection of fodder and green manure, and the grazing of livestock, part of the manure from which may be used on-farm.

We might also expect NGOs' concern with holistic approaches to poverty alleviation among the rural poor to give them a wider perception of agriculture than that of government in three further dimensions:

- If poverty is to be addressed adequately many problems identified in agricultural production require adjustments in the sequence of processes from input supply to processing and marketing. We might expect NGOs' flexibility to lead them to deal more fully with issues spilling over into these areas than would, for instance, a government department concerned solely with agricultural research;

- NGOs' concern with holistic and empowering approaches might, furthermore, be expected to lead them to address socio-economic and political obstacles to sustained technological change in small farmer agriculture. This would lead them to be concerned with the rural poor's security of access to land and credit, and with the expansion of popular and formal educational services in rural areas (cf. Figueroa and Bolliger 1986);

- In densely populated or highly inegalitarian societies, NGOs might be expected to search for and develop renewable natural resource based activities accessible to those women and men with little or no land, or to those who for cultural reasons do not normally work on the land.

The different aspects of these broad definitions of 'agriculture' and 'technology' are summarized in Box 3.1. The remainder of the chapter assesses how far, in what ways, and to what effect, NGOs do indeed operate with this broad perspective on ATD. To begin our assessment of how far the case study material supports these wide interpretations, we now examine briefly the range of material documented for this study, followed by in-depth focus on particularly illuminating case studies.

AN OVERVIEW AND ILLUSTRATIVE CASE STUDIES OF NGOs' WORK IN ATD

Table 3.1 provides a summary of the main characteristics – client groups, agroecological zones and technology focus – of the NGO case studies drawn from each region and documented in the companion volumes to this book.[3] Drawing on our discussion at the beginning of this chapter, it is evident that

---

Box 3.1 Broad-based definitions of agricultural technology, derived from expected NGO interpretations

---

- The production of annual and perennial crops;
- Livestock;
- Trees grown on-farm;
- The interaction among various combinations of these components (e.g. agroforestry, silvipastoralism);
- The maintenance and improvement of natural resources on-farm (e.g. soil and water conservation, genetic resource conservation);
- The relationship between on- and off-farm activities:
  - *sequentially*, e.g. seed production in the early part of the sequence, and crop processing in the latter part, and the relationship of each to crop production;
  - *spatially* e.g. on-farm production in the wider context of watershed management; the links through, for example, fodder and green manure between on- and off-farm production opportunities
- Activities particularly suitable for the landless or near-landless, and for women, such as kitchen gardens and backyard livestock production.

---

approximately one-third of case studies across the three regions based their actions on broadly production-oriented perspectives including, for crops, new varieties and (where possible) the corresponding agrochemical inputs and, for animals, health-care through conventional vaccinations. In Figure 3.1, we distinguish between two wider goal orientations within which these technological actions are expressed: the more orthodox production-oriented approaches which transfer technological packages into new environments without adequate consideration of local context, and the more grassroots-sensitive approaches, which draw down what is appropriate from the options available, and adapt it to local circumstances.

Production-oriented approaches

The focus of the present study has been on the more innovative NGO approaches to ATD and NGO–state interaction. It has not, therefore, documented in detail the numerous examples that exist in which NGOs simply transfer 'packages' of technology such as hybrid seed and agrochemicals to their clients without adequate consideration of their agroecological or socio-economic appropriateness. Several examples of this kind are, however, documented for Africa by Moris (1991) and by Smillie (1991).

None the less, evidence assembled by Kohl (1991) summarized in Box 3.2 indicates a propensity among certain supposedly progressive and innovative NGOs not only to introduce highly visible production-oriented technologies but to replicate them unquestioningly. This suggests weaknesses in their

Table 3.1 Approaches to agricultural technology development in case study NGOs

AFRICA

Orientation[1]	NGO	Client group(s)	Agroecological zone(s)	Technology focus
Production-oriented	CRS[2] (Gambia)	Small farmers via traditional groups (kafos)	Semi-arid/riverine	Varietal trials; small ruminants; sesame production and processing; vegetables; fruit trees.
	Action Aid (Gambia)	Small farmers	Semi-arid/riverine	Seed and input supply; extension, training, animal traction, horticulture.
	CUSO (Gambia)	Small farmers	Semi-arid/riverine	Varietal trials on major food crops; input supply; training.
	GSM (Gambia)	Small farmers	Semi-arid/riverine	Production and distribution of seed (generally minor food crops); input supply.
	TAAP (Ghana)	Small farmers	Guinea savannah	Input supply; extension; participatory trials, marketing and storage of maize.
	GVAM (Zambia)	Small farmers	Semi-arid lakeshore	Trials and demonstrations with grains, cowpea, sunflower, organic vegetables; animal traction.
	Silveira House (Zimbabwe)	Small farmers	Semi-arid plains	Dissemination of improved varieties of maize, sunflower, groundnut, with credit for fertilizer etc.
Agroecological	ACDEP (Ghana)	Small farmers	Guinea savannah	Agricultural information service; extension support unit; input supply; training.
	Langbensi (Ghana)	Small farmers	Guinea savannah	Input supply; animal traction; soil improvement; development of participatory methods; agroforestry; extension and training.
	Mazingira/CARE (Kenya)	Rural people	Various	Agroforestry; seed collection and seedling production.

KENGO (Kenya)	Rural people; other NGOs and GOs	Various	Research on indigenous trees; seedling production; extension and training.
Machakos Catholic Diocese (Kenya)	Rural people; other NGOs and GOs	Arid/semi-arid	Soil and water conservation within wider rural development strategies.
Rodale (Senegal)	Small farmers	Semi-arid	Soil and water conservation; integrated crop/tree/animal production techniques; information-exchange networking.
ARAF (Senegal)	Small farmers	Semi-arid plains	Soil fertility management (composting, manuring); use of leguminous crops; sheep fattening.
ORAP (Zimbabwe)	Small farmers	Semi-arid	Indigenous seeds research and distribution.
ENDA (Zimbabwe)	Small farmers	Semi-arid	Recovery of traditional seeds; seed networks; grain milling; community management of indigenous woodlands.

ASIA

Production-oriented	BRAC (Bangladesh)	Marginal farmers and landless; women	Central Bangladesh	Backyard poultry; landless irrigation and many others.
	Proshika (Bangladesh)	Marginal farmers and landless; women	Central Bangladesh	Backyard dairy; social forestry and many others.
	MCC (Bangladesh)	Small/marginal farmers	Central Bangladesh	Varietal testing and management practices in soya, vegetables, potato, rice.
	FIVDB (Bangladesh)	Landless; women	N.E. Bangladesh	Ducks: improved genetic material and management.
	PRADAN (India)	Mainly landless; some marginal farmers	Various; include floodplains and Deccan plateau	Raw silk; leather processing; mushrooms, among many others.

Table 3.1 Continued

Orientation[1]	NGO	Client group(s)	Agroecological zone(s)	Technology focus
	Ramakrishna Mission (India)	Small/marginal farmers	E. India coastal areas and floodplains	FSR; varietal trials and management practices in rice and vegetables; aquaculture; training of government staff.
	AKRSP (India)	Small/marginal farmers	Arid/semi-arid hills and coastal areas of W. India	Various technologies and management practices; development of participatory approaches to needs assessment, technology testing, and training by government staff.
	BAIF (India)	Small/medium farmers	Various, mainly central plains	Cross-breeding cows by artificial insemination; related animal health and nutrition; many other technologies.
	LP3ES (Indonesia)	Small farmers	Irrigated lowlands	Develop methods for farmer management of irrigation schemes; many other technologies.
Agroecological	CARE-Lift Project (Bangladesh)	Marginal farmers	Central Bangladesh	Biointensive vegetable production.
	RDRS (Bangladesh)	Marginal farmers; landless	Central Bangladesh	Treadle-operated low-lift irrigation pumps and many other mechanical technologies.
	AWS (India)	Small-scale farmers	Central plains	Integrated pest management and others.
	Auroville (India)	Own settlers (mainly foreign); small farmers	S.E. India coastal areas	Wasteland regeneration; watershed management; soil and water conservation; low-input management of traditional rice varieties.
	MYRADA (India)	Small/marginal farmers	Central plains	Options for watershed management; development of and training in participatory rural appraisal methods.

Organization	Target group	Zone	Activities
Nepal Agroforestry Foundation	Small/marginal farmers	Mid-hills	Nursery development for agroforestry species; related trials, demonstrations, training and networking.
United Mission to Nepal	Small/marginal farmers	Mid-hills	Forestry trials, demonstrations and training.
IUCN (Nepal)	Small/marginal farmers	Mid-hills	Participatory development of environmental plans at village and higher levels.
MBRLC (Philippines)	Small/marginal farmers	Hill farming	Integrated crop/tree/livestock practices for improved soil and water management on sloping land.
IIRR (Philippines)	Small/marginal farmers	Uplands	Agroforestry training materials, biointensive vegetable production and other technologies.
Mag'uugmad Foundation (Philippines)	Small/marginal farmers	Uplands	Diversification of farm enterprises introduced jointly with improved soil and water conservation methods.
ATA (Thailand)	Small farmers	Central and N.E. plains	Participatory adaptation of rice-fish farming and design of dissemination methods; research and networking on biological pesticides.

LATIN AMERICA

	Organization	Target group	Zone	Activities
Production-oriented	CIPCA (Bolivia)	Indian colonists	Lowland forest margins	Modern varieties, increased sowing density, machinery.
	AGRARIA (Chile)	*Mestizo campesinos* / *Mestizo* labourers	Semi-arid coastal brushlands; humid temperate lands	Modern varieties, fertilizers, vine cultivation, livestock vaccination, processing.
	SEPAS (Colombia)	*Mestizo campesinos*	Rainfed, semi-arid northern plains	Introduction of new crops, irrigation techniques, agroindustry.

Table 3.1 Continued

Orientation	NGO	Client group(s)	Agroecological zone(s)	Technology focus
	CESA (Ecuador)	Indian highland peasants	Rainfed Andean slopes	Modern varieties, limited use of agrochemicals, afforestation and limited terracing, processing.
		Lowland *mestizos*	Floodplain (the coastal plain)	
	GIA (Chile)	*Mestizo campesinos*	Rainfed semi-arid and humid	Farming system development, some mechanization.
	PROCADE (Bolivia)	*Mestizo campesinos*	Mid-altitude slopes	Modern varieties.
Agroecological	El Ceibo (Bolivia)	Indian colonists	Mid-slope forest margins	Organic production and processing of cocoa.
	CESA (Bolivia)	*Mestizo* colonists	Lowland forest margin	Agroforestry, legume winter crops and indigenous soil classification to manage slash/burn practices.
	FUNDAEC (Colombia)	Highland and valley *mestizos*	Rainfed mid-altitude slopes	Alternative systems, IPM, resource conservation.
	CAAP (Ecuador)	Highland Indian	Andes; rainfed slopes	Recovery and improvement of traditional rotation systems, native crops and native trees.
	IDEAS (Peru)	Highland Indian and *Mestizo*	Andes; rainfed slopes and plains; semi-arid coast	Diversified low external input systems and kitchen gardens; rural agroindustry.
	CIED (Peru)	Highland Indian and *Mestizo*	Andes; rainfed slopes and plains	Diversified low input and 'alternative' technologies; micro-drainage basin approaches.
	PROCADE (Bolivia)	Highland Indian	Andes; rainfed slopes	Agro-ecosystem development; low input technology.

Sources: Africa – Wellard and Copestake (1993); Asia – Farrington and Lewis (1993); Latin America – Bebbington and Thiele (1993)

Notes: 1 NGOs for each Region are divided into two broad groups according to whether their small-farm agriculture focus is production-orientated, principally via the use of external inputs, or whether their approaches are principally agro-ecological, with low (or zero) use of external inputs.

2 For full names of NGOs, see Glossary.

3 Indian refers to ethnically indigenous Amerindian groups; *Mestizo* to mixed race Indian-whites; *campesino* is the Spanish term for small poor farmers.

individual capacity for research, poor communication of experiences among them, and excessive haste to demonstrate impact.

Box 3.2 NGOs and the unquestioning replication of non-viable production-oriented technology – the example of protected horticultural systems in the Bolivian Andes

---

During the past decade some fifty Bolivian NGOs have introduced protected horticultural systems (PHS) in the high Andes (*altiplano*) in an attempt both to avoid climatological constraints and to meet nutritional needs. A survey of forty NGOs currently still involved in PHS was undertaken to examine why these projects have generally failed to meet their goals.

The principal findings are that few NGOs have conducted serious experimentation on PHS (and there is little available from public sector research on which they might draw); that the few conducting experimentation have not done so rigorously; that a 'folklore' of the supposed advantages of PHS has developed; that communication flows among NGOs regarding the outcome of PHS implementation has been inadequate, and that the rapid implementation encouraged by donors and by NGOs' own philosophy has led to premature introduction of unproven designs and management systems, and to a neglect of farmers' objectives and constraints.

In brief, a greenhouse for protected horticulture is highly visible to visiting donor teams (even from main roads) and so provides apparently convincing evidence that NGOs are making an impact. Its apparent ability to contribute to family nutrition is intuitively appealing, but closer consultation with rural people would have revealed priorities for activities capable of greater income generation, and doubts about whether the returns justified the effort involved in PHS. Finally, technical problems caused by high humidity and poor air-circulation were inadequately researched.

Greater information sharing among NGOs, which donors might stimulate, together with more systematic, farmer participatory approaches towards experimentation, could have helped to resolve these difficulties.

Source: Kohl 1991

Numerous cases have been documented, however, in which NGOs take more grassroots-sensitive approaches to modernization. Work by AGRARIA in Chile, for instance, has been concerned with the introduction of new crop varieties, inorganic fertilizers, and the vaccination of livestock (Aguirre and Namdar-Irani 1992); CESA has been concerned with similar issues in Ecuador (CESA 1991), and GIA, also in Chile, has, in particular circumstances, supported the introduction of mechanization.

Further examples raise wider issues of two kinds. First, contrary to some NGO rhetoric, production-oriented approaches in some circumstances are neither unsustainable nor biased towards higher income groups. Several examples of such instances can be drawn from NGOs' work on livestock: research by BAIF (Bharatiya Agro-Industries Foundation) on frozen semen technology for the promotion of cross-bred dairy cattle through artificial insemination (AI) (Satish and Farrington 1990, and Box 4.10 in this volume) led to much higher conception rates than earlier liquid semen technology. Equally importantly, BAIF's work demonstrated how AI delivery systems incorporating routine vaccinations, advice on health and nutrition and progeny monitoring could be established more cost-effectively than comparable services operated by government and in ways which – through the service charges levied – have some prospect of institutional sustainability (Satish and Kumar 1993). BAIF's work, however, is not geared towards the lowest income rural households. By contrast, work on ducks by FIVDB (Friends in Village Development – Bangladesh) (Nahas 1993) and on poultry by BRAC (Bangladesh Rural Advancement Committee) (Box 3.9) has succeeded in reaching poor women and the landless in Bangladesh. Delivery systems rely heavily on the services of local people (paid for by the NGOs' clients) and involve both improved genetic material and vaccines. So far they have functioned reliably and promise to be sustainable for the longer term. Similarly, work by BRAC (Mustafa *et al.* 1993) and by Proshika (Wood and Palmer-Jones 1990) on mechanisms for the ownership of irrigation pumps by groups of landless labourers has not only brought 'modern' technology within the reach of the poorest, but also used it to good effect.

A second set of issues relates to the particular dilemmas faced by those NGOs who genuinely seek to respond to farmers' requests, but then find that these are based on particularly short-term perspectives that seem obviously unsustainable. Thus, in its work with small farmers in Zimbabwe's communal areas, who were deprived of hybrid maize seed and fertilizer prior to independence, Silveira House made these technologies available on credit terms in response to farmer demand (McGarry 1993). Strong uptake by farmers encouraged the new government after independence to replicate the model, though with modifications which made it less successful (see Box 5.1). Concern that these technologies would be unsustainable unless accompanied by other changes led Silveira House to promote appropriate soil and water conservation methods among farmers, but the attempt was met by almost complete indifference. Early, vigorous and largely top-down promotion of conservation measures by NGOs would be seen by many as non-participatory and over-zealous – yet, if conservation measures are resisted until degradation becomes highly visible, restoration will become much more difficult to achieve, and a hands-off participatory approach may end up contributing to an unsustainable development.

Agroecological approaches

The steady emergence of an agroecological agenda owes a great deal to the accumulated experience and advocacy of many NGOs working with low input production systems. None the less, within this very broad perspective, there are different degrees of 'purity', and different motivations for using low input approaches. In Fig.3.1 we distinguished between ideologically-driven and more pragmatic responses to local contexts. We consider each in turn.

Ideologically-driven approaches

NGOs are not short on ideological motivation. Two of the areas in which they have been particularly active have been in the promotion of organic and environmental issues, and in the recovery and revalidation of ethnic cultures and their traditional technologies.

In Nepal, the approach to environmental planning at village and higher levels espoused by IUCN (International Union for the Conservation of Nature) is strongly underpinned by environmentalist convictions. Village Environmental Plans were drawn up in a pilot area by village-level committees constituted specifically for this purpose, working jointly with local NGOs. Particular attention was paid to the protection and promotion of those local institutions established, for instance, to manage forest resources. Efforts are in hand to integrate Village Plans with the development plans designed by government agencies for these areas. However, the 'vertical' character of government planning and the fact that it is driven by annual budget cycles are obstructing such initiatives (Carew-Reid and Oli 1993).

In Thailand, part of ATA's (Appropriate Technology Association) work is similarly driven by environmentalist concerns, themselves motivated by the appalling environmental health record in the country: in 1985, 5,500 persons, 384 of whom died, were admitted to hospital with pesticide poisoning; and toxic residues, principally of organochlorine insecticides, were found in 90 per cent of produce sampled by the Department of Agriculture in 1982–85. ATA has been instrumental in bringing together researchers in universities and government departments interested in working on biological pesticides, and in facilitating interaction between them and farmers so that traditional practices can be identified and built upon (Jonjuabsong and Hwai-Kham 1993).

Again, in the case of ENDA (Environment and Development Activities) in Zimbabwe, the search for ecological alternatives was driven by a rejection of the 'scientific agriculture' of the government, which had encouraged farmers to grow maize in areas more suited to drought tolerant crops. ENDA's Indigenous Seeds Project was established jointly with other NGOs and, more recently, with government departments, to identify, store and distribute local varieties of millets and sorghums which had traditionally been used in drought-prone areas (Box 3.3).

Farmers' local knowledge clearly pays a key role in the IUCN, ATA and ENDA examples cited above. However, some NGOs' actions are driven by a predisposition towards the strengthening not only of indigenous technical knowledge, but also of the social and cultural conditions in which it is rooted (cf. Bebbington 1991b).

Box 3.3 An NGO draws on and strengthens local knowledge systems – the case of ENDA in Zimbabwe

---

Research by Environment and Development Activities – Zimbabwe (ENDA) in villages located in the poorer agro-ecological zones, which experience recurrent droughts, revealed communities' need for seeds which are drought tolerant, early maturing, palatable, disease resistant, resistant to bird damage, store well and give a high yield with minimum use of expensive external inputs.

The drive by government and seed companies to promote 'scientific agriculture' together with changing preference towards maize which is easier to grow and process than small grains, has resulted in a serious loss of sorghum and millet cultivars.

In collaboration with the Ministry of Agriculture, and with other NGOs, ENDA embarked on an Indigenous Seeds Project and on seeds networking activities with the objective of facilitating the supply of the best available seed from the total gene pool, for specific (mainly dry) environments. The project was based on the assumptions that:

- Millets and sorghums have been grown by farmers in dry areas for many generations and are thus well adapted to the environment, but are under-researched;
- In pre-independence Zimbabwe no specific research was carried out for the communal areas so that, subsequently, the danger existed of introduction of inappropriate crops and varieties from higher rainfall areas;
- Efforts are necessary to preserve genetic material against natural degradation and against the risk of widespread consumption of seed during drought.

Networking among five NGOs facilitates the collection and testing of material from different agroecological zones. Around 200 varieties have been multiplied and screened and the project is now cleaning up seeds for wider distribution. Many of these have been collected and identified by farmers themselves.

Source: Chaguma and Gumbo (1993)

While the broadly empowering approaches taken by many NGOs (Chapter 4) involve some strengthening of local organizations, and so underline the principle that, to be sustainable, technological change in many instances must be supported by appropriate organizational and institutional change, it is rare that social and cultural conditions can be revalidated in the ways attempted by CAAP (Andean Centre for Popular Action) in Ecuador (Box 3.4). It is perhaps only in areas where ethnic, socio-political and economic divisions are clearly correlated that fervent commitments to such ethno-development approaches are found. Not surprisingly, they are thus encountered quite often in Andean America. No more than elements of this approach are found elsewhere as, for instance, in Mag-uugmad's efforts to use traditional Filipino forms of social organization (*alayon* work-groups) as a basis for its activities.

Box 3.4 Ethno-development approaches: an NGO seeks to restore and strengthen the social and cultural context in which indigenous knowledge is rooted: the case of Andean Centre for Popular Action (CAAP), Ecuador

---

CAAP was established in 1978. One of its objectives has been to identify the economic, socio-political and cultural conditions underlying the agricultural techniques and systems of indigenous Andean groups.

CAAP argues that *campesino* (peasant) production is based on forms of social organization and agronomic principles which 'modern' research should strengthen, rather than replace. In particular, *campesinos'* perceived objectives of long-term preservation of the ecosystem, the production of enough food to meet consumption and socio-cultural requirements, and the optimum use of available labour were seen as part of a structural context which should be protected and enhanced.

Initial efforts on CAAP's research farm – itself located in sloping, eroded conditions similar to those faced by *campesinos* – were not adopted by farmers since, despite CAAP's philosophy, they sought to demonstrate terracing and other soil conservation technologies hitherto unknown in the area.

Subsequent work concentrated on questions of productivity, pest control and the maintenance of soil fertility through new varieties, rotations and changes in cultural practices. Several years' data from trial plots are necessary before firm conclusions on the impact of changed cultural practices and rotations can be drawn. However, the beneficial effects of some rotations are already observable. By introducing different varieties from other locations through a small number of farmers who have opted to experiment with them CAAP has also succeeded in broadening the genetic base of local crops grown on-station.

Major remaining shortcomings are the lack of rigour in the design and conduct of experiments (especially of complex rotation trials), CAAP's limited efforts to exchange ideas and experiences with other research organizations, and internal debates over whether technology transfer is an appropriate function for an organization such as CAAP. On this last issue, the hitherto dominant faction has maintained that CAAP should focus on strengthening *campesino* knowledge and should allow *campesinos* to draw down what they see as potentially useful from the research station, and disseminate it through traditional channels.

Source: CAAP 1991; interviews

Pragmatically agroecological approaches

Our discussion of agroecological approaches has so far been concerned with those driven by predispositions towards organic or ethno-development perspectives. The majority of case studies exhibiting an agroecological approach are, however, inspired more by pragmatism than ideology. Several examples can be given. In India, Action for World Solidarity worked with government research and extension institutes and with universities to develop integrated management methods for a major pest of castor since the 'hairiness' of the caterpillar impeded penetration by conventional insecticides (Box 3.5). The Mindanao Baptist Rural Life Centre in the Philippines developed soil and water conservation technologies which, in addition to using practically no external inputs, fitted into labour availability profiles, and provided a regular income for farmers (Box 3.6). In Bolivia, El Ceibo designed and disseminated organic cocoa production technologies not out of ideological predisposition, but in response to two factors: the identification of a 'niche' in international markets which paid a premium for organic cocoa, and the high cost and limited effectiveness of chemical technologies (Box 3.7). As neo-liberal policies proceed apace, pushing up the cost of imported agrochemicals, we might expect such pragmatic adoption of low-input strategies to continue.

Box 3.5 Technical and organizational inputs in integrated pest management – the case of red-headed hairy caterpillar (RHC) (*Amsacta albistriga*) on castor in India

---

It is estimated in Andhra Pradesh that the damage caused by RHC to young castor seedlings causes crop losses of some 50 per cent. Small farmers are reluctant to pay the high prices of conventional insecticides, particularly since the caterpillar's 'hairiness' prevents penetration by droplets and so reduces their effectiveness. The search by Action for World Solidarity (AWS) for alternative, traditional control methods revealed some use of trap crops and manual collection of egg-clusters, both of which were of limited effectiveness. However, efforts to attract moths into bonfires as they emerged from the pupal stage appeared to offer promise. AWS therefore began to work on an integrated pest management system jointly with Andhra Pradesh Agricultural University, which provided better understanding of the life-cycle of RHC; the Institute of Plant Protection and Pest Surveillance which provided training for NGO and GRO staff in, for example, the interpretation of light trap results; the Technology Transfer Unit of the Zonal Coordination Office of the Indian Council for Agricultural Research, which arranged credit for the purchase of bonfire materials, and the supply of waste rubber for bonfires. As co-ordinating NGO, AWS worked with twelve GROs covering over 2,000 ha of farm land to ensure that the bonfires were lit simultaneously at peak emergence of the moths, and to co-ordinate activities building up to this. Much work remains to be done if this control method is to prove sustainable – for instance, trees will have to be planted locally to ensure the continuing supply of bonfire materials. None the less, early results, showing an increase in the value of castor production in 1989 of Rs4.8 million against a project cost of Rs1.5 million, are highly promising.

Source: Satish and Vardhan (1993)

Other patterns

The case study material reviewed also demonstrates several other significant patterns in NGOs' work with technology. Firstly, clear differences exist among the case studies in the *level* of approach. Some NGOs have an area focus to their efforts – thus, IUCN is concerned with area-based approaches to planning in order to reflect environmental concerns, and Myrada and CIED (Centre for Research, Education and Development – Peru) with 'bottom-up' approaches to watershed planning. At the opposite extreme, some are

---

Box 3.6 An NGO having a pragmatic agroecological approach which has been scaled up in the Philippines – the case of Mindanao Baptist Rural Life Centre (MBRLC)

The MBRLC, founded in 1971, aims through its Sloping Agricultural Land Technology programme to develop production and resource-management technologies capable of arresting and reversing the soil degradation visible in wide areas of hillside farming in Mindanao. Early efforts at gully-plugging on farmers' fields in response to their requests were unsuccessful, and the Agriculture Department was unable to provide technical guidance. MBRLC consequently established its own research farm and set up alley cropping trials with the leguminous tree, *leucaena*, which enhanced both maize yields and soil fertility. Livestock production, particularly dairy goats, were subsequently incorporated, followed by the incorporation of trees and tree crops of varying maturation periods. Integrated production systems for lowland farming and for vegetable production have also been developed.

Training at MBRLC has been provided for over 10,000 persons from NGOs, GOs and local farming communities. These and sales of planting material, constitute MBRLC's major source of income. In order to preserve its independence in a politically volatile area, it has a policy of refusing direct grants from government and from aid donors.

The detailed knowledge of local conditions on which MBRLC's efforts are based has resulted in options which are environmentally beneficial, and at the same time, respond to farmers' need for regular cash incomes while long-term measures for soil regeneration are being implemented. While sensitive to local knowledge, the technologies under trial are brought in from a wide range of government, non-governmental and international sources. While not predisposed towards organic approaches, the high costs and limited availability of agrochemicals has influenced MBRLC's work towards low external input technologies. The wide adoption of MBRLC technologies led to an invitation from government for MBRLC to participate in the planning and implementation of the major Southern Mindanao Agriculture Project (see Box 5.15), which was launched at MBRLC's headquarters in September 1990.

Source: Watson and Laquihon 1993

concerned with highly specific components of farming systems, such as ATA's work on biological pesticides, work by RDRS (Rangpur–Dinajpur Rural Service) on low-lift pumps and by El Ceibo on cocoa technology. The majority, however, fall between these two extremes.

Second, the extent to which NGO activities draw upon and strengthen indigenous technical knowledge (ITK) varies widely. A pattern can be detected which is broadly consistent with expectations derived from Figure 3.1. Thus, orthodox production-oriented NGOs in Kohl's view (Box 3.2) did not pursue sufficiently participatory approaches and sought to introduce technologies which lay well outside the realm of ITK. At the opposite extreme, ITK was a key determinant of the types of technology perceived by CAAP to be consistent with its ethno-development objectives. Also near this end of the

---

Box 3.7 Pragmatic agroecological approaches geared to market opportunities for organically grown cocoa – the case of El Ceibo in Bolivia

---

El Ceibo is a federation of thirty-six member co-operatives in the Andean valleys north of La Paz. Initially established to promote small farmers' (mainly cocoa growers') interests through input supply, credit, processing and marketing, in the early 1980s it began to undertake research and extension on cocoa and kitchen-garden crops as government services deteriorated rapidly. Early successes were obtained by El Ceibo's efforts to disseminate technologies developed by the NARS, specifically, pruning techniques against witches' broom disease. Particular factors underlying its subsequent success include:
- Clear client orientation in research, and strong mechanisms for consultation with, and feedback from, farmers;
- The training of extension agents and paratechnicians drawn from local communities; the training of paratechnicians at local universities;
- The strong financial base secured by El Ceibo for its research and technology transfer services, partly through a levy on cocoa sales, partly through foreign financial and technical assistance;
- The strong links developed by El Ceibo with cocoa research and producing organizations in other countries.

While trials on the response of cocoa yields to chemical fertilizer were initially part of El Ceibo's research portfolio, the emphasis in recent years has been on entirely organic methods for the production of cocoa, in order to take advantage of premium prices in foreign markets for organic products.

Source: Trujillo 1991

---

spectrum, work by AWS, ATA and ENDA sought to draw upon and strengthen local knowledge systems, which themselves were based on agroecological principles.

However, the great majority of case studies demonstrated lower levels of reliance on ITK. For most NGOs, the key ingredients of innovation are introduced from outside the locality, but local knowledge is important in determining in the first place the broad types of innovation that are likely to be successful and, subsequently, to a greater or lesser degree, in adapting innovations to suit local conditions.

Third, not all case studies fit squarely into the categories identified. Thus, for instance, Gwembe Valley Agriculture Mission in Zambia supports farmers' efforts to grow high-yielding varieties of maize and so can be classified as broadly 'production-oriented', but the high price and unreliable supply of horticultural inputs has encouraged it to experiment with organic vegetable production. Similarly, BAIF pursues broadly production-oriented cross-bred livestock strategies, but has generated innovations (including fodder production) through agroforestry which can be regarded as an 'agroecological' approach when compared with the dominantly orthodox approaches to technology in India.

Fourth, NGO approaches are not static: over time they may progress from one category to another. Thus, NGOs in northern Ghana (to cite but one example) were, until recently, largely concerned with the supply of conventional agrochemical inputs. However, a combination of increased environmental awareness and rising farm-gate prices of inputs led them to switch to low external input strategies. As we noted, this trend is increasingly apparent among NGOs.

Fifth, an additional component of the socio-economic dimension which cannot easily be incorporated into Figure 3.1 is the degree to which NGO approaches are market-oriented. It would be tempting, for instance, to suppose that production-oriented approaches reflect attempts to increase market orientation, that pragmatic agro-ecological approaches are less so, and that organic initiatives are primarily concerned with autarky. While this is a plausible framework for some of the case studies, important exceptions should be noted. It has not been lost on some that, where markets exist and are accessible, an organic product can be extremely profitable. Profit is arguably the main motivation for El Ceibo's efforts to reach international markets with organically-grown cocoa (Box 3.7).

Finally, it is clear from Table 3.1 that NGOs' work is indeed concentrated in the adaptive sub-system of the ATS. Even the pathbreaking work of BAIF, whose structure of laboratories, research station and field-level monitoring is superior not only to that of any other NGO in the sample, but also to many government research services, essentially involved the adaptation of known artificial insemination technology to Indian conditions.

CHARACTERISTICS OF NGOs' 'ALTERNATIVE' APPROACHES TO ATD

Discussion so far in this chapter has focused on NGOs' approaches to ATD within broad production-oriented and agroecological schools. Agroecological approaches and those which we have termed 'grassroots-sensitive' production-oriented share the philosophy of drawing down from a range of technological options those most appropriate to local opportunities and constraints. They generally involve higher levels of participation of the rural poor in the choice of technology and may be said to represent an approach to development which is *alternative* to 'technological package' and 'blueprint' production-oriented approaches.

Among those NGOs pursuing some sort of development alternative, several aspects of their approaches are commonly encountered. These include:

1 A focus on social organization as an integral component of ATD;
2 Methodological innovativeness and experimentation;
3 A fusion of research and extension functions;

4 The design of change within the sequence of activities in the agricultural cycle;

5 A concern for linking on- and off-farm resource management and recognition of local diversity;

6 A focus on technologies for the rural landless and women.

1 The focus on social organization

Much of the technology emerging from orthodox production-oriented approaches is transferred to individual farmers rather than to groups of farmers. Although notable exceptions exist, few extension agents are either trained or provided with the facilities to engage in group approaches.[4] By contrast, as we discuss in Chapter 4, working with groups is central to the mode of operation of many NGOs. There are many reasons for this. In part it is attributable to the fact that their concepts of participation and empowerment are more easily developed among groups than with individuals. It is also indicative, in some cases, of the socio-political backgrounds of many NGOs in collectivist and anti-individualist social thought. Finally, many NGOs assert that if they are to achieve wide and sustained dissemination, some technologies are better introduced through group than through individual approaches. For example:

- In Bangladesh, BRAC (Box 3.9) and Proshika (Khan *et al.* 1993) succeeded where government had failed in reducing disease and mortality among improved livestock (cattle, poultry) through the creation of sustainable systems relying heavily on interaction among local people for the delivery of vaccines and, in the case of poultry, cross-bred chicks. Similar results were achieved for ducks by FIVDB (Nahas 1993) where, in addition, the NGO had itself conducted a breeding programme in an area largely neglected by government. While the access to irrigation pumps by landless labourers facilitated by BRAC and Proshika was clearly dependent on access to credit and technical skills, another a key component was the organization of a cohesive group around each pump capable of working together to obtain optimum irrigation performance, to meet credit repayments and to negotiate water supply contracts with farmers (Wood and Palmer-Jones 1990; Mustafa *et al.* 1993);

- In the local-level production of seed, a combination of social organizational and technical inputs is similarly important. Henderson and Singh (1990) report how NGOs in The Gambia and Sudan organized farmers' groups capable of producing seed which maintained its characteristics and was of adequate quality. Such organizational efforts are particularly important where the variety to be produced is open-pollinated and so must be grown at a distance from other plantings of the same crop. CESA in Ecuador promoted the production of improved seed potatoes within traditional systems which relied on a high degree of farmer organisation.

Organization of this type was necessary in order: (1) to manage the production of seed at higher altitudes where incidence of disease was lower, allowing trade of these seed potatoes for potatoes intended for consumption produced at lower altitudes; and (2) to monitor the production and distribution of seed among the farmers who were members of the organization. Numerous further examples of NGOs' efforts, not all of them successful, to combine organizational and physical inputs in seed production are provided by Cromwell and Wiggins (1993);

- In integrated pest management, a high degree of social organization is again required to ensure that maximum effect is achieved through the simultaneous implementation of certain activities. The joint efforts of an NGO in India (Action for World Solidarity) together with grassroots organizations and government research and extension services illustrate the complex conditions that have to be met for success (Box 3.5);

- While part of Silveira House's effort focused on the provision of improved varieties, credit and agrochemicals, a key factor in the success of its efforts was the formation of groups to guarantee loan repayments. Loans given at the beginning of the season were repaid after harvest was sold to the marketing board. Only organized groups were allowed to register with the marketing board. They also agreed to the deduction of the loan repayment by the marketing board, and its transfer to Silveira House. Repayment rates were much lower when this ingredient was omitted by government's large-scale replication of Silveira House's work (McGarry 1993);

- In Indonesia, work by LP3ES to organize groups of farmers and to discuss and meet their concerns was a key factor in achieving the successful handing over of certain irrigation maintenance and operating responsibilities to them. LP3ES assisted in ensuring that certain practices within small irrigated areas subsequently subsumed into larger ones, were respected. It also provided training to irrigation staff in social and institutional issues, and joined a working group with government to investigate how handing over might take place. This combination of initiatives by LP3ES produced results which government had previously been unable to achieve on a sustained basis (Bruns and Soelaiman 1993).

2 NGOs, methodological innovations and experimentation

Many NGO efforts, whether in 'production-oriented' or 'agroecological' contexts, have been underpinned by the tenet that detailed consultation with clients is essential if their perceptions of opportunities and constraints, their local technical knowledge and the socio-economic contexts in which they work are to be brought adequately to bear on the agenda for research and on the identification of relevant innovations.

The methodological innovations that NGOs have introduced, in working with both modern and agroecological technologies, have been based largely on these perceptions. Two examples from the wide range available in the case studies are particularly illuminating:

- In India, MYRADA (Box 3.8) has been promoting participatory rural appraisal (PRA) methods among government services and other NGOs (Fernandez and Mascarenhas 1993), drawing largely on its own experience in watershed management (Bhat and Satish 1993);
- In Chile, GIA's field presence and its contacts through research with institutes in other countries facilitated its development of farming systems methodologies which were subsequently drawn on in the restructuring of the NARS following the return of democratic rule (Sotomayor 1991).

3 A fusion of research and extension functions

Government research and extension services are characterized not only by a clear distinction between these two functions, but, in most cases, by their location in separate departments. A further institutional subdivision is found in some countries where applied and adaptive research services fall under one department, and basic/strategic research services under a different one. In the Philippines, for instance, university research into agricultural issues falls under and is partly funded by the Ministry of Education, and in India, the agricultural universities and extension services are responsible to state governments, whereas a large number of agricultural research institutes fall under the national government.

Recent research argues that these institutional divisions hinder effective interaction among the respective functions, exacerbating the lack of client orientation in research, and hampering effective feedback of farmers' views on the technologies which have been disseminated (Merrill-Sands and Kaimowitz 1991; Kaimowitz 1991).

As Table 3.1 indicated, all the NGO work in ATD reviewed here lies at the adaptive end of the research spectrum. This may, in itself, facilitate stronger links with clients than are generally found in NARS, but a number of other reasons also suggest themselves:

- Differences in motivation and reward systems between researchers working in NGOs and those in NARS (as discussed in Chapter 2);
- Few NGOs have separate departments responsible for research and for dissemination. In many cases, these are the responsibilities of different *individuals*, but institutional arrangements – often aided (and demanded) by the small overall size of the organization – are generally such that those responsible for disseminating and monitoring the progress of a technology are in close contact with those responsible for its design. Thus, the rural

---

Box 3.8 NGOs and the development of new methods – the case of MYRADA and participatory rural appraisal in India

---

MYRADA has been involved in rural development since 1968 and now works in some 2,000 villages, primarily in S. India. In recent years, it has worked to promote and develop pilot approaches for people's participation in the planning and implementation of watershed development. The experience it gained in the PIDOW Watershed Development Project in Karnataka State (Bhat and Satish 1993) encouraged it to promote participatory rural appraisal methods among individuals and other NGOs. By mid-1991, it had trained over 3,000 persons in these methods. A Workshop held in Bangalore in 1991 was an important step in bringing together thirty participants from over ten NGOs and five representatives of government departments who had been developing PRA techniques (Mascarenhas *et al.* 1991).

Reports at the Workshop from senior staff in Karnataka State Watershed Development Cell confirmed that PRA had contributed to villagers' self-confidence in interacting with planners, taking on responsibility for the monitoring of construction and development activities conducted by the state, and themselves undertaking some construction activities from local resources.

However, a number of difficulties remain, including:

- People's continuing lack of identification with structures designed and constructed by government agencies, and consequent difficulties of ensuring adequate maintenance;
- Continuing problems of top-down attitudes among government staff, including a reluctance to search out issues that affect villagers, expecting, instead, that villagers will wait on them;
- Continuing expectations among villagers that government officials will perform poorly and will demand bribes.

Source: Fernandez and Mascarenhas 1993

---

development officers responsible for implementing BAIF's artificial insemination technology have clear channels of communication with researchers, including regular meetings, and researchers are keen to make use of the data they provide on, for example, progeny testing (Satish and Farrington 1990; see also Box 4.10);

- Barriers between research and dissemination are also minimized in NGOs by the fact that their work is not generally organized by disciplines or commodities, but is projector issue-oriented. The NGO imperative is towards *action* on the issues identified. While generally effective in focusing effort, this action orientation can sometimes be over-emphasized. In the case of Bolivian NGOs, a desire to show visible impact (largely inspired by a concern to show donors results and so secure continued funding) resulted in the dissemination of technologies that have not been adequately tested (Kohl 1991; see also Box 3.2). NGOs working in the FITT programme in The Gambia exhibited some of the same shortcomings (Gilbert pers. comm.; see also Box 5.11).

4 NGOs and the design of change within the sequence of activities in the agricultural cycle

Most NARS are concerned with research and extension only during the production phase of the agricultural cycle. Prior activities such as input supply and credit are commonly handled by other agencies, both governmental and private sector, as are post-harvest processing and marketing operations. The larger the number of different agencies involved in these pre- and post-production activities, the greater the likelihood that technological change and research in one stage of the food system will be out of cycle or inconsistent with the requirements of other changes.

We do not claim that NGOs have been uniformly successful in meeting clients' requirements across phases of the production cycle, but there are at least indications that issue-oriented approaches, together with the flexibility of non-hierarchical management structures, have generated integrated approaches in some cases (cf. Carroll 1992). Thus, PRADAN's work on improved leather processing (Box 5.21) involved them in working together with a local group on the construction of a processing plant, the acquisition of inputs and of credit, the identification of suitable training courses and the establishment of sales and feedback links with the private commercial sector. In a less complex example, MCC in Bangladesh realized that soya could not become a successful small farm crop unless new market opportunities were created. It thus set about exploring snack food markets and processing requirements with the private commercial sector (Buckland and Graham 1990; see also Box 5.4). Again, the provision of processing facilities was a key factor in the early success of sesame introduced into the Gambia by CRS (Owens 1993). In El Ceibo, the move into ATD activities was consequent upon earlier cocoa processing work. Market contacts led El Ceibo to realize the importance of improving production technology, and in the absence of effective public services El Ceibo initiated its own ATD. Similarly, the move into organic technologies has been a response to El Ceibo's market experience which has helped highlight organic cocoa as a lucrative niche market (Box 3.8; Trujillo 1991).

5 NGOs and interactions between on- and off-farm resource management

The approaches taken by most NARS – even those permeated to some degree by farming systems perspectives – are restricted to issues falling within the farm boundary. Only in exceptional cases (e.g. Lumle and Pakhribas Agriculture Centres in Nepal – Farrington and Mathema 1991) is research into on- and off-farm resource management integrated, as for example, in the case of research into the fodder and green manuring properties of trees grown off-farm. Yet, as Jodha (1986) argues, a high proportion of the income of poor households is made up from biomass acquired from common property or open access areas.[5]

An awareness of the importance of off-farm biomass as an income source for the rural poor is found in some of the NGO case studies reviewed here. Thus, for instance, AKRSP has been pressing for joint management by the rural poor of forest reserves in India (Sethna and Shah 1993 – see also Box 4.8), and, in some cases, NGOs seek to integrate research across on- and off-farm natural resource management. Similarly, in the Cajamarca area of Peru, CIED is seeking collaboration with public sector agencies to draw up a development plan for the area, which will contain coherent proposals for technical research in both on- and off-farm areas (Guerrero 1991). More generally in Peru, NGOs have for some time promoted a micro-drainage basin approach to resource management and development planning that would engage both on- and off-farm resources. This approach has a West African parallel in the 'village land use management' approaches taken up by NGOs (and GOs) (e.g. Sumberg 1991).

6 NGOs and technologies for the rural landless and for women

NARS see *farmers* as their clients, so the technologies they develop may have little relevance to the landless or near-landless. Natural resources-based technologies relevant to these, if they are catered for at all, are generally under the mandate of ministries charged with rural enterprise, not with agriculture.

Analysis of the case studies confirms the observation by Carroll (1992), that few NGOs are concerned with the creation of employment opportunities suitable for the rural poor. Highly innovative exceptions are found in some densely-populated countries. The work of Proshika and BRAC in Bangladesh in pioneering schemes by which small groups of landless labourers own and operate irrigation pumps has already been mentioned. Such schemes have involved BRAC in the provision of credit and training in technical and managerial skills to the landless. They represent an opportunity not only for employment among the landless, but also for them to participate in the expanding market for input supply to those components of agriculture undergoing rapid modernization. In Chile, AGRARIA trains landless youths in pruning skills in order to enhance their prospects of employment in vineyards (Aguirre pers. comm.).

Finally, Box 3.9 indicates the complex social organizational arrangements which BRAC, together with local people, had to design in order to facilitate the uptake of backyard poultry by women and the landless.

Overall, it is tempting to classify many of these innovations which cut across production-oriented and agroecological approaches as inputs which are missing from government approaches and which NGOs provide. Silveira House's provision of high-yielding varieties of maize, fertilizer and credit to small farmers in Zimbabwe (McGarry 1993), or CUSO's (Canadian Universities Service Overseas) provision of fisheries specialists as a catalyst for work on rice-fish farming in what had hitherto been a Thai *agricultural* research institute (Sollows *et al.* 1993) seem to be examples of such a gap-filling exercise. Certainly, the visible products of NGOs' work in many cases appear to fill gaps left by government. But the idea that NGOs provide 'missing inputs' is by no means a complete assessment of what NGOs do, nor why they do it. As we have argued, this work is rooted in their wider economic, political and social objectives which, as we shall see in Chapters 4 and 5, also underpin the types of interaction in which they engage with their clients, and with governments.

Box 3.9 Technologies for women and the landless: improved poultry production promoted by the Bangladesh Rural Advancement Committee

---

In Bangladesh, almost 50 per cent of rural households are landless or near-landless, and women face cultural restrictions on work outside the household compound. Livestock production is one activity which can be conducted within the compound. Poultry production alone is estimated to account for 23 per cent of per capita animal protein consumption in the form of both meat and eggs, but mortality is high and productivity low.

Following a number of unsuccessful efforts to upgrade poultry production in 1979–83, by the mid-1980s, BRAC has devised a complementary set of technical and local institutional innovations which by 1990 had been replicated by government and other NGOs in 7,400 villages, affecting some 10 per cent of poultry production. The innovations comprised:

- One poultry worker (female) per 1,000 birds, trained in rearing techniques, health care and vaccination;
- Vaccines for the poultry worker provided by the Department of Livestock (DoL), and training provided jointly by DoL and BRAC; her remuneration covered largely from vaccination fees;
- The establishment of systems to allow poultry keepers access either to day-old chicks from government breeding farms or, if they felt confident enough to handle such young birds, to two-month-old chicks reared at special units set up by BRAC and DoL, but managed by local key rearers, and capable of rearing batches of 250 chicks from day-old to two months;
- A feed production centre serving several villages to provide a balanced feed supplement for cross-bred stock;

Scaling-up of the scheme means that over 33,000 key rearers are now operating commercially, and almost 5,500 poultry workers have been trained. Demand for day-old chicks from government hatcheries has risen from 0.5 million/year in the mid-1980s to almost two million currently. However, the system remains crucially dependent on the capacity of government to deliver inputs, especially vaccines, down to local level.

Source: Mustafa *et al.* 1993

CONCLUSIONS

This chapter has reviewed NGOs' experiences in agricultural technology development, and in general supports the assertions that:

1 NGOs take a wider view of 'agriculture' and 'technology' than do government research and extension services. This view embraces:

(a) Interactions among crop, livestock and tree components of agriculture;

(b) The full range of activities in the agricultural cycle, from input supply to processing and marketing;

(c) The wider conditions (e.g. of land tenure, credit availability and education) which must be met if agricultural change is to be sustainable;

(d) Interactions between on- and off-farm resource management; and

(e) The types of resource management opportunity accessible to women and the landless.

This is not to imply that all of these perspectives are shared by all NGOs: some inevitably focus their efforts more narrowly than others. This diversity of views and experience among NGOs inevitably influences the prospects of collaboration between NGOs and NARS, or even of one side learning from the other. This is treated in Chapter 5.

2 NGOs' efforts to promote technical change in agriculture fall into two broad groups: production-oriented approaches, in their more orthodox form, are premised on the transfer of packages of technology and of support systems with little sensitivity to local context and operate with the broad goal of transforming traditional agriculture (cf. Schultz 1964). In the case of crops, such packages comprise combinations of high-yielding varieties, agrochemicals, mechanization and improved water supply. By contrast, agroecological approaches focus on systems interactions between crops, livestock and trees, low external inputs and technologies deriving from, or capable of reinforcing, indigenous knowledge systems. While the more opportunist NGOs may adopt one or other of these approaches without much reflection, for many they are the result of serious deliberation. Whether to adopt production-oriented or agroecological approaches, and what alternatives within these, is for many a result not only of technological feasibility and financial considerations, but of their views on the broad technological approaches appropriate to particular societies under specific circumstances. It is therefore as much a political as a technical decision.

3 Sensitivity to local context varies widely among NGOs, but strong field presence gives some a higher awareness of local conditions than found in government. This awareness feeds into the ways in which technological change is implemented in practice. Thus, efforts to 'modernize' need not be restricted to the orthodox approaches outlined above: from the range of modern technologies available, grassroots-sensitive approaches draw down those relevant to local needs and seek to devise support systems compatible with local socio-economic, cultural and institutional conditions. A wide

range of agroecological approaches is also noted: some NGOs' actions are driven by their own or donors' ideological commitments to, for instance, organic agriculture, environmental conservation or ethno-development. For others, agroecological approaches are selected on pragmatic grounds in response to local constraints and opportunities.

The type and level of interaction with indigenous knowledge systems and with the market proved to be two further important dimensions for the classification of NGOs' work on agricultural technology: predictably enough, close integration with input and output markets was of greatest importance with production-oriented approaches, but a negligible feature of ethno-development. Some NGOs attached much more importance than others to market opportunities, and to the scope for sustainable market-oriented institutional forms to be built up among their client groups. Further evidence on this is examined in the discussion in Chapter 5 of links with the private commercial sector. While many NGOs drew on indigenous knowledge in establishing what broad types of innovation would be acceptable, the more important sources on which they drew for innovation almost invariably lay outside local communities and, in many cases were located in modern scientific knowledge spheres. Many NGOs sought to strengthen local capabilities to absorb, experiment with and modify such innovations, but only in ethno-development cases was the attempt made to revalidate and strengthen the social and cultural contexts in which indigenous knowledge is rooted.

The empirical material suggests other features of NGOs' work on technical change in agriculture which apply across production-oriented and agroecological approaches, through at different levels and in different ways. They have a strong focus on: the development of participatory methods of problem diagnosis, and technology testing, dissemination and evaluation; the creation of social organizational forms necessary for the sustainable implementation of technology; 'issue'-oriented approaches in which barriers between research, dissemination and feedback are minimized; and the need to relieve constraints not only in agricultural production technology, but also in the wider contexts of processing, marketing and input supply. The work of some NGOs explores interactions between on- and off-farm management and, in some contexts of high population density, certain NGOs have developed technologies particularly suited to women and the landless.

The evidence suggests that some innovations devised or promoted by NGOs – such as FIVDB's work on ducks, BRAC on poultry, Proshika with livestock, BAIF with cattle improvement, MBRLC with techniques for farming on sloping land and ATA with rice-fish farming – have had wide economic impact. However, comparisons of costs and benefits could be made only in a few cases (AWS, BAIF), and only in the case of BAIF was it possible to draw comparisons between the cost-effectiveness of NARS and of NGOs. Limited evidence in these areas is attributable to the facts that:

- Benefit:cost comparisons cannot easily be made because of the exploratory nature of most NGO efforts, and because of the high proportion of intended qualitative benefits, such as enhancement of local knowledge or capacity for experimentation among farmers. It is only recently that efforts have been made to address these methodological difficulties (Riddell and Robinson 1993);
- For the same reason, NGO–NARS comparisons of cost-effectiveness can rarely be made because the objectives of each do not coincide exactly: NGOs' objectives tend to be broader and less easily quantified;
- Monitoring and evaluation capacity is under-resourced on both sides.

Further questions of impact are treated in Chapter 4, especially in relation to the empowerment of the rural poor, and in Chapter 5 regarding their influence on the ATD agenda of NARS. However, it is already clear that most NGOs emphasise the goals of *sustainable* impacts in both agroecological and socio-economic senses, and the establishment of strong local institutions capable of enhancing the technical capabilities of the rural poor and strengthening their position *vis-à-vis* the public sector, other social interests and (in many cases) the market.

We wish to conclude this chapter on a note of caution: discussion of the wide range of innovative activities undertaken by NGOs may have generated the impression that they are poised to make a major impact on the livelihoods of the rural poor through agricultural change. To draw such a conclusion would be highly misleading: as we stress elsewhere, ours is not a random sample: case studies were deliberately drawn from the larger and, probably, the more innovative NGOs. Furthermore, even in countries such as Bangladesh, which has a particularly high density of NGOs, it has been estimated (Lewis 1993) that even if there were no overlap among NGO efforts, only 20 per cent of the rural poor would be reached. In practice, as we discussed in Chapter 1, NGOs' efforts are poorly co-ordinated both among themselves and between NGOs and government. Suffice it here to refer back to the discussion of NGOs' weaknesses in Chapter 1 – small size, inadequate access to specialist skills and facilities, and a preoccupation with local issues which distracts attention from the main policy debates – all of which imply a limited impact for NGOs unless their ideas on technology, methods and relations with clients are taken up by government. This issue is central to the discussion of interaction between NGOs and the state in Chapter 5.

NOTES

1 If it were not such a clumsy expression 'renewable natural resources management' should ideally be used through this book to reflect NGOs' perspectives.
2 N. S. Jodha (1986) indicates that for parts of Andhra Pradesh in India, up to 25 per cent of gross farm income among poorer households is made up from off-farm natural resource exploitation.

3 Bebbington and Thiele 1993; Farrington and Lewis 1993; Wellard and Copestake 1993.
4 These exceptions are often linked to strongly donor-influenced projects – such as, for instance, the Agricultural Technology Improvement Project in Botswana (Norman and Modiakgotla 1990).
5 Exploitation of the resource is managed in the former case, but not in the latter.

4

NGOs AND THE RURAL POOR: PATRONAGE, PARTNERSHIP OR PARTICIPATION?

INTRODUCTION

Much of the reputation of NGOs is based on claims about their ability to reach the poor. Reaching the poor, is not, however, the same as alleviating poverty (and far less eradicating it). In considering the extent to which NGOs' work with agricultural technology has any impact on rural poverty, we must therefore subject them to a doubly critical appraisal. First, we need to ask how far they do in fact reach the poor, and whether their relationships with rural people are as good as is claimed. Second we need to address how far their work addresses the causes of rural poverty – as opposed simply to treating some of its symptoms. To the extent that their work is wanting in this second respect, then the questions arise as to how far changes in government action and policy may be necessary, and how far NGOs may be able to foster such changes. These questions are dealt with in Chapters 5 and 6.

Before we discuss our empirical material, we therefore wish to place it in the context of three interrelated discussions. We first discuss recent observations on the extent, location and dynamics of rural poverty. We then consider two recent evaluations of NGOs' work with the poor, both of which have suggested that NGOs rarely get to grips with the underlying causes of poverty. This discussion leads into a brief review of different perspectives on the role of NGOs in rural poverty alleviation, which also suggests that much of the contemporary interest in NGOs similarly emphasizes a role for them which will not address many of the underlying causes of rural poverty.

These opening sections provide the context against which to assess this study's empirical material on NGOs' poverty reach in the development of agricultural technology. This is the theme of the major section of the chapter. We examine NGOs' performance in the various strategies through which they seek to work with the rural poor. Much has been written about NGOs' work in participation and empowerment of the poor, and therefore we make a particular effort to analyse the different meanings attached to these concepts. We examine how, and how far, they can be applied to the agricultural technology development work of NGOs and government research and extension services. In essence, this discussion comes to conclusions similar to those in the two other studies reviewed.

Before closing the chapter with a summary of the main conclusions, we examine wider issues arising from the discussion. The first is the question of how far development organizations actually give space to the rural poor in setting the development agenda. The second is the scope for modest, but wide-scale, improvements in the relationships between government organizations and their clients: this is frequently overlooked by those intent on more radical, but often idealized approaches. This scope is, however, limited by the wider implications for the state of empowerment among the rural poor.

The generally sanguine conclusions of this chapter have similar implications to those of Chapter 3 on the limits of NGOs' technical capacity: namely that neither NGOs nor their new fans must underestimate the importance of government and policy.

IDENTIFYING THE RURAL POOR AND THE CAUSES OF THEIR POVERTY

While identifying the rural poor might be simpler than identifying the causes of their poverty, neither is an easy task (IFAD 1992).

There are diverse ways of measuring rural poverty. The highly aggregated measures of poverty found in national and international statistics can, for instance, be couched in both relative and absolute terms. Relative measures designate as poor a certain percentage within a given population – typically the bottom 5 or 10 per cent of the national or regional distribution of income. In contrast, absolute assessments need to identify a standard and fixed measure that can be used for across-country comparisons – this is methodologically much more problematic. Early absolute measures were based on a per capita dollar-equivalent GDP (Gross Domestic Product) figure set at the same level across a number of countries. Inconsistencies in the methods of estimating GDP and differences in relating monetary measures to the widely differing levels of goods and services necessary to keep individuals above the poverty line in different countries, led to the increasing use of physical measures, such as those based on calorie intake. These, however, were again problematic because for instance, of the difficulty in assessing the intra-household distribution of food.

Given this diversity of measures, it is not surprising that recent global estimates range from 630 million 'extremely poor' people, through 780 million 'poorest of the poor' to 1225 million 'living in absolute poverty' (Kates and Haarman 1992; see also Box 4.1). Furthermore, within these measures we might distinguish between those who are 'poor' in relation to general consumption standards, and the 'ultra poor' who differ from the poor in their attitudes towards risk and in their capacity to respond to income-generating opportunities (Lipton 1988).

A great deal of this poverty is rural. Mellor (1988), for instance, suggests that poverty is primarily a rural phenomenon and is concentrated in 'low potential' rural areas. In this he echoes Chambers's (1983; 1987) assertion that the poor are particularly concentrated in complex, diverse and risk-prone environments.[1]

Box 4.1 Regional distribution of poverty and projected trends

Estimates in 1990 suggested that over 1 billion people live in poverty, of whom two-thirds are found in Asia and one-quarter in Africa. Population growth means that, despite a reduction in the percentage of people living in poverty between 1970 and 1990, the poor increased in absolute numbers. Aggregate improvement has been achieved in many poverty indicators over the last thirty years: per capita calorie intake has risen by some 20 per cent; infant mortality fell from 232 deaths per 1,000 live births in 1960 to 112 in 1990; life expectancy at birth rose from forty-six years in 1960 to sixty-three in 1987, and adult literacy from 46 per cent to 64 per cent in 1970–85.

However, these aggregates conceal wide variations in performance: in particular, many of these trends were weak or even negative in Sub-Saharan Africa, where per capita incomes fell by almost 15 per cent during the 1980s. On current trends, 400 million Africans are projected to be living in absolute poverty by 1995, against 290 million in 1990. By contrast, South Asia is expected to see a decline from 530 million in 1985 to some 510 million by 2000, and East Asia from 280 million to 70 million over the same period.

Source: World Bank 1992

Such spatial correlations should not be read directly as explanations of poverty, since within rural areas different types of poverty can be identified. In both high and low potential areas there is, for instance, what might be called 'interstitial poverty', an effect of the relations and processes creating landlessness and low wages; similarly in both sorts of environment there may be 'overcrowding poverty' (a result of pressures on resource bases leading to severe subdivision of assets) and the 'traumatic' poverty that comes in times of crisis, particularly war (IFAD 1992).

However, low potential and vulnerable areas are, many agree, particularly prone to poverty aggravating processes (IFAD 1992; Turner and Benjamin 1991; Leach and Mearns 1992). Furthermore, these vulnerable areas (and people) seem more likely to experience self-perpetuating downward spirals of poverty. A recent review of thirty case studies correlating poverty and environmental degradation demonstrated surprising similarities emerged in the mechanisms through which the poor lost resource *entitlements* (Kates and Haarman 1992; Sen 1981). Reduced entitlements exacerbate livelihood vulnerability which, in turn, increases the pressure to 'mine' natural resources (cf. Chambers 1987). The authors go on to identify three types of positive feedback between poverty and degradation which aggravate this resource 'mining': (1) displacement of populations leading to increased population in the receiving area and the subdivision of finite resources; (2) rapid endogenous population growth leading to the same outcomes; and (3) insufficient resource bases for people to be able to maintain the protective works, or social relationships, that help them cope with natural hazards (cf. Turner and Benjamin 1991).

These analyses echo Bernstein's (1977) arguments about the simple reproduction squeeze, which draws attention to the ways in which the dynamics of market relationships (for Bernstein the dynamics of capital) lead farmers to adopt unsustainable resource use practices, 'rationally, and sometimes rationally in desperation' (Chambers 1987). They also link our understanding of the dynamics of poverty to Goodman and Redclift's (1991) recent reflections on the politics of sustainability, which in addition emphasize the ways in which resource conflicts between poor and rich similarly aggravate poverty. These conflicts occur both locally (as in struggles over land and water) and nationally (in the competition to control the policy environment).

Many of these analyses have emphasized above all the deleterious effects of rapid population growth. Now, population increase need not lead to resource degradation, and in cases where the poor have, for instance, access to product and capital markets, secure tenure, and freedom from violence, then demographic increase can go hand in hand with agricultural intensification (Turner and Brush 1987; Tiffen and Mortimore 1992). More often than not, however, these favourable contexts are missing, and poverty and degradation each result from rapid demographic increase (Turner and Benjamin 1991). This makes the data in Table 4.1 particularly relevant for our assessment of the impact of the case study NGOs. These data indicate that, although some countries covered in our study already face declining population pressures in agriculture, for others, particularly in Africa and S. Asia, population pressures will continue to grow for several decades.

The foregoing is a very brief review of a massive literature, but it is the basis for making several points of relevance to our thinking about the poverty impacts of NGOs.

First, there are very many processes, relationships and socio-economic structures underlying rural peoples' poverty: landlessness, low wages, political powerlessness, occupancy of vulnerable biophysical environments, imperfect markets, adverse macroeconomic policy environments, and elite-controlled policy processes are but a few. To expect NGOs to address all (or any) of these is naive; but for an NGO, one implication is that if the organization is to have a sustainable impact on poverty, then at least some of these underlying causes must receive systematic attention.

Second, the more important of these diverse causes of poverty cannot be identified a priori. The precise processes by which people become and remain poor are specific to socio-economic and agro-ecological contexts. To devise

Table 4.1 Agricultural population projections in the case study countries

	Turning point[1]	Total population in agriculture (000) 2025	Percentage of total population in agriculture 2025
ASIA			
Bangladesh	2010–20	93,618	43
India	Beyond 2025	621,052	51
Indonesia	1985–90	51,532	19
Nepal	Beyond 2025	29,040	86
Philippines	2010–20	30,433	30
Thailand	2000–10	30,090	35
LATIN AMERICA			
Bolivia	Beyond 2025	4,380	24
Chile	1960–70	878	5
Colombia	1985–90	5,317	10
Ecuador	1990–95	2,335	10
Peru	2010–20	8,257	20
AFRICA			
Gambia	Beyond 2025	1,004	67
Ghana	Beyond 2025	14,193	30
Kenya	Beyond 2025	48,925	59
Senegal	Beyond 2025	12,504	70
Zambia	Beyond 2025	12,444	52
Zimbabwe	Beyond 2025	16,493	51

Source: United Nations 1988.
Note: 1 i.e. period in which annual rate of growth of labour force in agriculture is nearest to zero.

appropriate interventions therefore requires detailed local knowledge of the type NGOs claim to possess.

A related implication of these conceptual and policy frameworks is that poverty is expressed in different ways in different parts of the world and that therefore the groups on which NGOs focus will also necessarily vary. For instance, many NGOs in South Asia concentrate their efforts not on farmers but on the landless. Conversely, in those African countries where land is less of a constraint, NGOs have targeted households who may have access to arable land but lack appropriate or adequate farming technologies such as ploughs and draught power. Many of these low income households may be female-headed and face exclusion from formal agricultural extension efforts as a result of gender bias.

Finally, it may be argued that some of the dynamics of poverty are such that to address them adequately necessarily requires government intervention. Similarly, government not only has a key role in defining the policy framework, it also shares with NGOs and private commercial agencies the responsibility for implementing many of the required strategies. If NGOs are to address poverty in a sustainable fashion they will therefore have to influence government policy and action.

NGO PERFORMANCE IN REACHING THE POOR: RECENT EVIDENCE

In a recent review of the relationships between poverty and environmental degradation, Leach and Mearns (1992) insist that rural employment gene- ration is essential if the poor are to cease mining their resource bases (Leach and Mearns 1992; cf. Bebbington 1992). A similar review for Latin America suggests that the link between rural poverty and environmental degradation will only be broken once policy biases against sustainable resource use strategies among the rural poor are removed (de Janvry and Garcia 1992). How do NGOs measure up to these and other challenges?

While there is evidence of NGOs' comparatively strong performance in welfare or relief activities, there is less evidence of success in income and employment generation. Clark (1991: 54) has argued that, despite their rhetoric, NGOs find it difficult to assist the growing numbers of people with few or no assets via income-generating projects. Tendler (1987) has come to similar conclusions, and suggests that many NGO interventions do not reach the bottom four deciles of the income distribution spectrum. These claims are also in large measure substantiated by some very recent studies of NGO performance in working with the poor – although it should be stressed that these studies also point to very many NGO strengths. It is helpful to draw lessons from the results of two of these: a review of thirty indigenous NGOs in Latin America (Carroll 1992), and another of sixteen Northern NGO projects in Africa and Asia (Riddell and Robinson 1993).[2]

Carroll's book grows out of an evaluation of NGOs supported by the Inter- American Foundation, or IAF (Carroll 1992). Seven of these were membership service organizations (MSOs) and twenty-three were grassroots service orga- nizations (GSOs). In general, most of these NGOs 'tried to achieve tangible benefits by supplying services that are needed or requested by beneficiaries, and most support beneficiary groups . . . not only [as] a matter of efficiency in the scale of services, but also [as] a commitment to collective empowerment as an independent value' (ibid, p.26).

Carroll evaluated these organizations against a range of criteria, which he grouped under three broad headings:

1 *Development services*: how far did NGOs deliver services (such as agricultural extension) appropriate to beneficiaries' needs, how far did these services lead into other positive changes, and how far did they reach the very poorest?
2 *Participation and empowerment*: how far were NGOs responsive and accountable to beneficiaries; and how far did NGOs reinforce base (or

grassroots) capacity, that is, the capacity of a group to 'create new systems and mechanisms to accomplish its goals'? (ibid.: 33). In Carroll's view, building such capacity is 'the heart of participation';

3 *Wider impact*: how far have NGOs produced innovative ways of solving problems which have been, or have the potential to be, 'scaled up' by other organizations? Also under this heading was an assessment of how far NGOs have succeeded in influencing policy, or have the potential to do so.

The assessment of the organizations against these criteria was subjective, but based on in depth and inside knowledge of the different projects. This scoring produced the results shown in Table 4.2.

Table 4.2 The performance of a sample of 30 Latin American NGOs

	Distribution of scores		
Criteria	High	Medium	Low
Service delivery			
Service effectiveness	24	6	0
Poverty reach	12	18	0
Participation			
Responsiveness/accountability	12	13	5
Reinforcing base capacity	12	8	10
Wider impact			
Innovation	13	12	5
Impact on policy (actual)	6	11	13
Impact on policy (potential)	15	13	2

Source: Adapted from Carroll 1992: 35.

The generally high scores in service delivery and participation reflect the fact that Carroll's sample was not random – these were organizations that had already been selected on these criteria in order to receive IAF funding in the first place. However, other aspects of the results are particularly interesting. Carroll's finding that membership support organizations (MSOs) perform less well than GSOs in practically all activities is important. To his mind it reflects the fact that despite their formal mechanisms for accountability, MSOs' responsiveness to members' needs is limited by their remoteness from base co-ops, and by the tendency of minority interests to dominate these membership organizations.

Carroll also concluded that NGOs' capacity to benefit the poorest groups in rural areas was limited. He (1992: 67–8) concluded that: 'even the highest rated [NGOs] have relatively few direct beneficiaries among the poorest of rural households'. He identifies several reasons for this:

• The widespread view among NGOs that the landless can only be assisted by (in the short term) employment generation programmes or (in the long term) land reform. This attitude, and the associated implication that such programmes are beyond the capacities of NGOs, reinforces their preference to work with small, semi-commercial farmers having some land;

- 'Self-selection' by those individuals participating in NGO programmes, which tends to bring forward those who are more experienced (or project-wise), active, and willing to take risks.

The implication is that while these NGOs may be very strong in promoting a participatory and indeed an empowering form of development, the people participating and being empowered are not necessarily the poorest – though they are doubtless poor. Carroll's purpose in stressing these points is not to reject NGOs (he is clearly impressed by them) but rather to de-mystify them. The challenge, then, is to find the sort of development strategy in which the poor can benefit from spillover effects, even if they do not participate directly in the projects involved (cf. Tendler *et al.* 1988).

A second study, by Riddell and Robinson (1993), comes to startlingly similar conclusions. In an assessment of the impact of sixteen income-generating projects of British NGOs and their local counterparts in Bangladesh, India, Uganda and Zimbabwe, the authors conclude that twelve of the sixteen projects broadly achieved their objectives and had a positive impact on poverty alleviation. Only two were judged to have failed against both criteria, and a further two achieved partial success. However, within this success were latent weaknesses similar to those identified by Carroll.

While highlighting the importance of beneficiary participation in the design and subsequent modification of projects, and the quality of NGO leadership and staffing as factors contributing to positive outcomes, the study notes that, in these as in Carroll's Latin American cases, many of the projects failed to reach the very poorest. Where they did, these successes were characterized by close 'targeting' of identifiable groups of the poor, and attempts to generate benefits among the poor that would then 'trickle down' to the poorest (through the stimulation of economic activity and new opportunities). However, where improvement in economic status did take place, it was modest, and there was little evidence that many beneficiaries had really managed to break out of the sorts of self-reproducing spirals of impoverishment about which Kates and Haarman (1992) and Chambers (1987) talk. In some cases, the sustainability of impacts was undermined by adverse physical environments. In others, the NGO had paid insufficient attention to what Carroll would call strengthening base capacity in local membership organizations, to give them the capability of carrying forward the work after NGO withdrawal.

Riddell and Robinson also question another article of popular NGO-speak – namely that NGOs achieve their impacts at a low cost. In their case studies, they suggest that the costs of many NGO interventions are high in relation to the benefits achieved. The costs per beneficiary were indeed comparable with those in government projects – though the benefits were generally higher.[3] Whilst some of these costs are incurred in group formation and awareness creation, and so may be regarded as a long-term investment, in other cases (e.g. group credit) they relate to the inherent administrative demands of the intervention. Of the sixteen projects examined, five were clearly not sustainable financially, while seven exhibited limited potential for financial sustainability.

The similarities between the findings of the two studies are striking. They are also important, for they suggest that most of the NGOs they dealt with, both national and international, found it difficult to break into and reverse the processes that create poverty.

Another point on which the studies concur, is that NGOs not only had trouble achieving this at the local level at which they were operating: they were also weak in going beyond the locality and achieving 'wider impact' through changing policy or strategy in the public sector. Riddell and Robinson comment that most of the projects 'repeatedly underplay the significance of economic and social factors outside the immediate parameters of individual projects' (ibid: Ch. 5). In Carroll's (1992: 35) evaluation NGOs' impact on policy had been very modest. Interestingly he suggested that the NGOs most able to have such impact were sub-national NGOs operating at a provincial level, and in several localities. Smaller and more locally-based NGOs are too small to have an impact on policy and cannot absorb the costs involved in deliberating with government. Conversely, efforts of NGOs in the capital cities to influence government institutions and policy are fraught with the difficulties stemming from the inflexibility and politicization of central government. Conversely, at a sub-national level, NGOs are large enough to absorb the costs involved in working with government, interpersonal ties are more likely to cut across NGO-government boundaries and have a positive impact on policy discussion, and the NGOs are more likely to command respect. The implication is that, to the extent that government decentralizes functions and decisions, NGO potential for influencing development policy may grow.

PERCEIVING A POVERTY ALLEVIATING ROLE
FOR NGOs

Judith Tendler (1982) has suggested that most NGOs feel that the defining characteristic of their approach is the ability to construct high quality relationships with low income people. Though she was writing a decade ago, she could have said the same today. Most NGOs still claim that they are more able than is government to reach the poor, and explain this primarily in terms of their more 'participatory' and hence empowering approach to development actions (Clark 1991; Fowler 1990). The findings of Carroll, and Riddell and

Robinson would suggest that NGOs are correct in this self-characterization. Furthermore, NGOs might justly insist that such participatory approaches can have poverty alleviating effects. They might claim that strengthening local membership organizations, and building base capacity, will contribute to the rupture of the poverty cycle. The greater the administrative and political strength of these organizations, the greater the likelihood that they will be able to influence regional political and economic processes. The social transformation that many NGOs are seeking to affect is deemed in NGO-lore to derive directly from an empowered rural poor who will identify the roots of their poverty, and then change them (Korten 1990). In this vision, achieving the wider policy changes necessary for sustainable poverty alleviation is therefore an indirect, because laudably participatory, process.

This is a reasonable assertion – and yet it may not be so straightforward. If most NGO projects have few direct, positive and self-sustaining impacts on the income of the rural poor, then one might ask whether NGOs can have any significant empowering impact. If, as Jonathon Fox (1990a) suggests, poverty and high risk livelihoods are two of the most significant obstacles to popular participation in membership organizations, then the *sine qua non* of political empowerment must be economic advancement and security.

This suggests that in their rhetoric NGOs may, to use a British idiom, have put the cart before the horse: in other words, they have started where they ought to have finished. By emphasizing that the changes necessary for sustainable poverty alleviation would result from political action of the poor they have inverted means and outcome. If the organizations of the rural poor remain weak as long as their members' livelihoods are insecure, those organizations will only have the strength to tackle the causes of their poverty once those livelihoods and incomes are strengthened. It is not casual coincidence that the strongest membership organizations are often those that have been concerned above all with supporting production and self-sustaining economic enterprises (Bebbington *et al.* 1993; Trujillo 1991).

If there is some truth in these (slightly mischievous) observations, then NGOs' strongest impact, at least in the short term, may be achieved by effective service delivery, that is, in the language of some observers, by operating as an extension of the welfare state. NGOs may have identified participatory methods through which to identify local needs and therefore make their service more effective than the welfare states. None the less, as in the experience of the welfare state, this service delivery has rarely become a self-sustaining, popularly managed process, because to the extent that NGOs have been unable to create sustainable income opportunities for the poor, they have not given the poor the means through which to escape their dependence on the services the NGOs provide them with.

One implication of these observations is that NGOs might be well advised to pay more attention to devising viable income generating strategies, before, or at the same time as, working with popular educational materials. Similarly, they might have more impact on poverty by seeking to change policy environments – in those cases where there is sufficient political freedom to do this (which are by no means all cases).

A second implication is that the prominent donor and government view of NGOs (namely that they are well equipped to fill the gaps left by both state and market in poverty oriented service provision) is perhaps an accurate reflection of what NGOs have in fact done (Box 4.2). Of course, this is not the image that many NGOs, in particular the more progressive ones, would have wished to convey (Bebbington and Farrington 1993). However, it is perhaps the image that bears closer resemblance to the reality of NGO projects to date – the reality of their effect, if not of their intention.

This slightly contentious commentary is not intended to reject all that NGOs have done. In the following pages we point to many cases of positive impact. It is, however, to suggest that there has been a tendency to over-emphasise the softer aspects of social development at the expense of developing economically viable income generating opportunities, and of engaging more vigorously in scaling up and having wider policy impacts. Similarly, we recognize there have been good reasons for this emphasis: NGOs have aimed to compensate for the state's dominantly economic view of development, and in other cases have avoided contact with repressive regimes. Ultimately, however, it is a strategy that with difficulty will have a sustainable impact. We turn now to discuss these points with reference to the material from this study.

REACHING THE POOR IN AGRICULTURAL TECHNOLOGY DEVELOPMENT – EVIDENCE FROM THE CASE STUDIES

Our discussion now turns to the mechanisms and strategies which the case study NGOs have used in attempting to reach the rural poor in the context of agricultural technology development. These embrace approaches which are *participatory* at various levels, and a number of more broadly *empowering* approaches.

It is worth stressing from the outset that many NGOs in their relations with people stress the importance of *processes* of joint experimentation and reflection rather than discrete *projects*. For instance, writing of one of the case study NGOs (Proshika), Wood and Palmer-Jones note:

> While projects are designed between groups and the field staff with as much forethought as possible, new forms of social action obviously generate unforeseen processes and problems, which have to be studied by those involved *as part of the social action itself*.
>
> (Wood and Palmer-Jones 1990: 25)

In acknowledgement of the importance attached by NGOs to 'process' approaches, the discussion of practical problem solving that follows is not

---

Box 4.2. Current multilateral agency views on strategies for poverty alleviation and on the roles of various types of implementing institution

---

Recent documents on poverty issues by the World Bank (1990b; 1991a, b) and UN agencies (UNDP 1990) stress the complex causes of poverty and the multiple strategies required for its alleviation. These include combinations of:

- Wider provision of social services (basic health, primary education) primarily by the state;
- The setting up of effective safety nets, to guard against sudden shocks and guarantee food security. Responsibility for these can be shared among the state, NGOs and commercial agencies;
- The targeting of any subsidies or welfare transfers specifically towards the poor;
- Labour-intensive economic growth designed to provide both employment and self-managed income-generating opportunities. Responsibility for this can be shared among public and private sector agencies.

These documents recognize that to escape from poverty, the poor seek resilient and sustainable livelihoods, and that a primary function of development policy is to support them in this aim. The state has a key role to play both in specific initiatives, but, equally importantly, in setting out an enabling policy frame for efficient production and equitable distribution within which NGOs and private commercial organizations can operate. While overall responsibility lies with the state, the most effective strategies against poverty are perceived to require effort from a multiplicity of institutional types, ideally working in complementary ways.

---

couched exclusively in terms of specific projects. Rather we discuss it as the development of a capacity among the rural poor to solve problems for themselves.

Participatory approaches

NGOs have long realized that participation by intended beneficiaries in the design, implementation and evaluation of development activities can enhance their effectiveness. More recently, major funding agencies, reviewing the performance of projects they have assisted, have come to share this view. Paul (1991: 2), quoting from the World Bank's experience, acknowledges 'the important role that the poor themselves can play in initiatives designed to assist them'. As examples, of this, he cites the enhanced success that was achieved by World Bank sponsored agricultural development projects in Sub-Saharan Africa once women farmers had played a role in their design, or farmers' organizations had participated in the maintenance and operation of irrigation, or group credit schemes had been organized.

It is clear, however, that wide variations exist in the ways in which participation is interpreted and applied. Our first task is to set out a simple framework for considering the types of participation observed in the case studies; we then discuss some of the wider issues raised by the evidence.

In an early attempt to analyse participation in ATD, Farrington and Martin (1988) broke participation down into that occurring at different stages of the sequence of activities during which technology is developed: problem diagnosis, technology testing and dissemination. Focusing on the diagnosis and testing stages of on-farm research, Biggs (1989) in turn analyses farmer participation in terms of its intensity. He identifies four levels, or degrees of participation in ascending order of intensity (Biggs, 1989: 3):

contractual – In which researchers merely hire inputs (land, labour) from the farmer, but make little effort to seek her or his opinions;

consultative – In which the farmers' opinions are actively sought by the researcher, who then develops the solutions;

collaborative – In which the researcher and farmer are partners in the research process;

collegiate – In which farmers and researchers interact as equals, and researchers aim to strengthen farmers' own informal research activities.

This classification suggests that *depth* of interaction is an important dimension by which participation can be classified.

Biggs's analysis was based largely on the work of public sector research and extension services whose mandate does not normally extend beyond technology development and dissemination. Participation, even a degree of empowerment, can be achieved within this narrow thematic context as farmers begin to conceive of new ways of engaging with researchers. However, most NGOs seek to raise rural people's awareness of the problems and opportunities they face in a much wider livelihood context, and then to pursue such empowering options as, for instance, the formation of claimant groups and wider social mobilization to address specific issues, which may or may not relate to agricultural technology, as they arise. NGOs' work in participation and empowerment needs therefore to be considered not only for its depth, but also in a second dimension, that is, of the *scope* of subject matter.

These two dimensions, depth and scope, are presented in Figure 4.1 as intersecting axes in a simple quadrant diagram – a framework which is derived from case study evidence, but appears consistent with the dimensions of depth and scope discussed above. What we call shallow participation with a narrow subject-matter focus (the upper left quadrant) is characterized by activities which address a limited range of farm activities and give clients (who are usually contacted as individuals rather than groups) little scope to influence experiments relating to these activities and limited opportunities to feed their reactions into further research on a continuing basis. The organization pursuing such shallow participation would essentially expect its mode of

103

SUBJECT MATTER SCOPE
Narrow

- Focus on a limited range of farm activities[1]
- Limited scope for clients to influence experiment
- Preference for individual over group approaches
- Limited feedback of results into research service
- Role for NGO continues unchanged

- Focus on a limited range of farm activities
- Lengthy interaction with clients to ensure that options for experimentation meet their needs, and that they 'own' the process of change
- Enhanced capacity of clients to interact upstream (e.g. input supply) and downstream (e.g. processing) with public and private commercial sector in support of the technical options
- NGO passes increasing share of management responsibilities to the clients

DEPTH Shallow   Profound

- Broad focus on a range of development options, in and beyond agricultural change
- NGO interaction with local groups is limited to, e.g., using them as a source of ideas
- Role for NGO continues unchanged

- Broad focus on a range of development options, in and beyond agricultural change
- NGOs work with groups in long-term conscientization and empowering modes; work on agricultural change emerges from this process and is set in the context of other economic and societal change
- NGOs work to develop links among local groups and create a capacity to draw on public and private sector services
- NGO passes increasing share of management responsibilities to the clients

Wide

Figure 4.1 The depth and scope of NGO approaches to participation ATD

Note: 1 The term 'activity' refers to both agricultural and non-agricultural activities, and to production and processing

operation to continue unchanged into the future. The remaining three quadrants summarize the characteristics of other combinations of depth and scope.

Two patterns in Figure 4.1 merit emphasis. First, deeper levels of participation tend to rely more on group than on individual approaches. This raises a number of questions regarding group formation which we address below. Second, the stronger the genuine desire of NGOs to hand over most if not all of their functions to self-sustaining membership groups, the more empowering is the approach. Boxes 4.3 to 4.6 provide examples from the case study material corresponding to each of the quadrants in Figure 4.1. While the contrasts among characteristics of the four quadrants are drawn out particularly starkly by the selection of extreme examples, in reality each axis represents a spectrum of possibilities and most of the case studies we reviewed occupy less extreme positions.

Box 4.3 NGOs at shallow levels of participation with narrow subject matter scope – the case of the Good Seed Mission (GSM) in The Gambia

---

GSM was established in 1985 by a former Peace Corps Volunteer who has been in The Gambia since 1967. It aims 'to feed the hungry body and soul' through a combination of improved agricultural practices and Christian teaching. Contact with farmers was built up as they came to the Mission for assistance with agricultural inputs, advice and repairs to equipment.

The production of improved seed on the Mission's own farm has been a major focus of agricultural activity. GSM participated actively in the FITT (Farmer Innovation and Technology Testing) programme (Box 5.11) and continues to attend monthly district-level extension planning meetings. Under FITT, GSM selected varieties on the basis of its own perceptions of what would constitute viable and useful additions to local farming systems. Most testing was done on GSM's own land, but farmers in four villages were invited to test improved cowpea varieties on their own fields, and improved cassava varieties in one village. These trials were conducted under farmers' practices, with modifications to take into account the principal recommended management requirements of the varieties. Farmers' assessments were used by GSM in determining whether to proceed with dissemination of the varieties.

Source: Sarch 1993, Wiggins, pers. comm.

Empowering approaches

The shift in some branches of development theory during the 1980s away from the prescriptions of top-down change towards an alternative development model has at its root a conception of empowerment as a form of social change brought about by local problem-solving techniques. The process is

---

Box 4.4 NGOs at deep levels of participation with narrow subject matter scope –
the case of CESA (Ecuador)

CESA (Ecuador) works with 150 grassroots organizations having over 8,000 member-families in testing and disseminating technologies which offer a middle way between idealistic ethno-development approaches relying on the revival of traditional techniques, and the high-input, costly, risky and potentially unsustainable technologies of the Green Revolution. Longstanding contacts with local communities provide detailed information from farmers on potentially suitable technologies.

In addition, CESA aims to foster the emergence of federations linking the communities. The goal is that these federations take on administrative functions, including liaison with the CESA itself, and negotiations with state agencies. These federations have played an important role in organizing the testing, multiplication and distribution of new seed and other technology among communities. The federations also interact with other NGOs and with government and private commercial agencies, exerting a demand-pull in relation to their agricultural requirements and, increasingly, in relation to wider rural development needs.

Source: CESA 1991 (in Bebbington and Thiele 1993)

Box 4.5 NGOs at shallow levels of participation with wide subject matter scope –
the case of the Ramakrishna Mission (RKM) in India

RKM was established in 1897 near Calcutta, initially as a religious organization, but has provided broad support for a number of welfare and development activities in the last forty years. RKM currently has some 250 staff based in 1,500 villages in West Bengal who help to design and implement social welfare and rural development activities. RKM's approach has been to support village-level youth clubs as a catalytic link between villagers and external agencies. Groups comprising approximately twenty-five youth clubs constitute a Cluster Club, and a district-level Youth Council interacts with the government's district administration. Youth clubs meet for one week each year with RKM to share experiences, and discuss options for future development.

Youth clubs receive orientation and training from RKM staff, and conduct 'needs surveys' in order to guide RKM's activities. They also help to identify other development activities in the villages, whether financed externally or relying on internal resources, and monitor the extent to which RKM's activities achieve immediate goals, and contribute to sustainable development in the longer term.

Source: Chakraborty *et al.* 1993

'. . . nurturing, liberating even energizing to the unaffluent and the unpowerful' (Black 1991: 21).

In some agencies, this has led to a shift away from emphasizing the

---

Box 4.6 NGOs in deep levels of participation with wide subject matter scope – the case of the Organization of Rural Associations for Progress (ORAP) in Zimbabwe

---

ORAP was established in the early 1980s by teachers, students, ministers of religion and women's groups in Matabeleland and Morlands provinces to promote understanding among rural people of the forces contributing to poverty, and of the options for enhancing livelihoods. Emphasis is on developing the ability of rural people to assert their right to acquire outside resources that would contribute to their chosen options, and to reject 'development interventions' that would not.

ORAP's structure has evolved to allow members to work together within and across groups on chosen tasks. Based on traditional *amalima* groups, the *unit* of ten or so families shares certain agriculture-related tasks; the *group* of five to ten units takes on larger projects such as the construction of a dam; an *umbrella* of five to ten groups is a forum in which group representatives co-ordinate activities among themselves and with ward-level government officers, and an *association* performs similar tasks at district level.

ORAP provides some general services (e.g. transport), and specialist equipment and staff support the wide range of activities chosen at the various levels, including education, health, agriculture and rural industry. A major focus of the agriculture programme has been to discourage the use of hybrid maize seed supplied by government in low rainfall areas where its performance is uncertain and instead to encourage the collection and use of traditional varieties of more drought-resistant millets and sorghums by setting up its own storage, multiplication and certification facilities. ORAP draws selectively on government extension services for technical advice and training, and has implemented a number of development projects with external funding obtained through them.

Source: Ndiweni 1993

---

completion of particular projects, toward greater attention to the importance of the development process itself (Clark 1990). However, like participation, the concept of empowerment resonates with a powerful component of rhetoric and has entered the current discourse of development with a multiplicity of meanings.

Friedmann (1992: vii) places empowerment at centre stage in his discussion of the politics of 'alternative development': for him an empowering approach to development is one that places the emphasis on enhancing decision-making autonomy, local self-reliance (but not autarky), direct (participatory) democracy, and experiential social learning. Its starting point is the locality, because civil society is most readily mobilized around local issues.

In an ideal world, empowerment would not be achieved through top down structures or by being imposed by outsiders. Rather it would be brought about through people's own involvement in development, as they learn increasingly to articulate their needs and their established rights. However, given the constraints encountered by the poor in organizing themselves, there remains a valid role for outsiders in strengthening and, at times, creating groups in response to identified needs in the context of a broader empowerment strategy. The role of external agencies such as NGOs is perceived as that of facilitator – assisting the emergence of capabilities, where invited to do so.

Below, we examine case study material against a number of the processes generally associated with empowerment, especially in relation to working with client *groups*.

Group versus individual approaches

As Figure 4.1 suggested, group approaches are more conducive than individual approaches to deeper levels of participation and hence to empowerment. Some of the case study NGOs have consciously moved from an individual to a group approach in the search for fuller participation. Thus, Langbensi agriculture station in Ghana abandoned its individual 'contact farmer' approach to extension in favour of group approaches when it found that individuals were keeping some of the best ideas to themselves (Kolbilla and Wellard 1993).

Working with groups, however, poses a number of problems for NGOs, some of which are well-documented in the wider literature. The formation of new groups is time-consuming. Hence some agencies are inclined to work with existing groups (e.g. RKM – cf. Box 4.5). The experience of Mag-uugmad Foundation Inc (MFI) in the uplands of Cebu, Philippines (Cerna and Miclat-Teves 1993) demonstrates that this can be done successfully: MFI worked with farmers to develop soil and water conservation technologies within the context of existing *alayon* reciprocal work groups.[4]

However, in many cases, existing groups are likely to be dominated by elites; they also tend to be characterized by the predominance of men over women, and by the landed over the landless.

Creating new groups does not, however, necessarily solve the problem of elite domination. On rare occasions, for example in newly-settled areas where elites have not yet succeeded in controlling local politics (such as those in Pakistan in which AKRSP (Aga Khan Rural Support Programme) works (Khan 1992), NGOs have the relatively easy job of forming community-wide groups which are not prone to domination by existing elites. On the whole, however, notions of harmonious communities in which all work together towards shared goals are being increasingly criticized (Scoones and Thompson 1992; Fairhead 1990), so that the very use of the term 'community' requires extreme caution. Experience from one of the case studies (TAAP – Tamale Archdiocese Agriculture Programme – in Ghana), for instance, highlights problems of ethnic conflict within some community-based groups (Millar 1991).

For these reasons, NGOs frequently find it desirable to identify or encourage the establishment of new groups, often on a 'special interest' basis within existing communities. This forms the basis, for instance, of BRAC's work with backyard poultry producers (Box 3.9) and, as the example suggests, is a particularly important device for enhancing women's development prospects. A further example of work with very poor 'interest' groups is provided by PRADAN's efforts (Box 5.21) to build upon caste-identity among flayers' groups in India, in order to enhance the quality of leather produced.

Advocacy on behalf of the rural poor

NGOs have sought to represent sections of the local rural population in their efforts to address wider constraints on development. This may involve taking up issues at the local level with agricultural extension agencies or legal institutions, or at the national level with central government. These efforts are intended to have 'wider impact' in Carroll's terms.

Partly influenced by the experience of Proshika's social forestry programme in Bangladesh (Box 4.7), Paul notes that

> Interactions arising from service-delivery roles tend to be relatively passive, but the activist, advocacy, and protest roles of NGOs can be expected to be stormier and more acrimonious.
>
> (Paul 1991: 7)

For these reasons, an NGO wishing to influence the state through joint practical effort can ill-afford to strike self-righteous postures at every opportunity, particularly, as Bruns and Soelaiman (1993) note, in countries such as Indonesia, where relations between NGOs and government are still fragile. Some evidence of NGO specialization in response to this reality appears to be emerging: in the Philippines, for instance, IIRR (International Institute for Rural Reconstruction) works closely with a number of GOs and so eschews advocacy, whereas the PRRM (Philippines Rural Reconstruction Movement) views working links with government as less important to the achievement of its goals, and so does not hesitate to challenge government on, for example, issues of land reform.

Two examples of NGOs aiming to represent rural peoples' interests directly through advocacy are cited here – the work of Proshika to defend villagers' rights to forest access in Bangladesh (Box 4.7) and that of AKRSP in pressing the Indian Forest Department (Box 4.8) to allow joint management of forests between villagers and officials, with a facilitating role for NGOs.

Elements of advocacy are, however, also evident in some NGO work which is primarily focused on agricultural technology development, including the work of ENDA on small grains in Zimbabwe (Chaguma and Gumbo 1993; see also Box 3.3) and that of the Latin American Consortium for Agroecology and Development (CLADES) (Bebbington and Thiele 1993).

CLADES, a consortium of Latin American NGOs committed to

Box 4.7 Advocacy in social forestry – Proshika in Bangladesh

The Bangladesh NGO Proshika has worked with groups of low income forest dwellers whose access to the *sal* forests in Tangail District (on which they depend for their livelihoods) is being resisted by local elites. The elites, in alliance with government officials of the Forest Department, have gradually shifted policy towards non-sustainable exploitation of the scarce forest lands for short-term gain. Proshika has therefore helped to organize and support local groups in their local level activities to secure better access and secure recognition of their rights under the law, as well as carrying out advocacy on their behalf at higher levels of government. Although some significant gains have been made, the encounter highlights the potentially violent conflicts generated by head-on confrontation between poor farmers and the local power structure, and the contradictions between national level resource conservation policies and local commercial interests.

Source: Khan *et al.* 1993

Box 4.8 The Aga Khan Rural Support Project (AKRSP): influencing wasteland development policy in India

---

AKRSP formed a loose coalition in 1988 with two other NGOs in an attempt to change government procedures which, since 1980, had increasingly restricted village access to government-owned land, so depriving them of the minor forest products and timber thinnings which contributed to their livelihoods. AKRSP's involvement in forest management by villagers began with a 17-hectare plot made available by sympathetic officials. However, despite the project's success, access to further land was denied by senior Forest Department officials at both state and national levels. There began a series of meetings with officials in which a key role was played by an NGO leader who had formerly been a senior civil servant. Field visits by officials to the existing site were also arranged, and the NGOs made use of their contacts to encourage publication of a number of national newspaper articles broadly supportive of the NGOs' position.

Persistent pressure from the NGOs over a three-year period culminated in government agreement that villagers could have usufruct rights to forests providing that agricultural crops were not grown in them. This concession was embodied in policy instructions issued by the Ministry of Environment and Forests on 1 June 1990, though considerable work remains to be done to secure their implementation.

Source: Sethna and Shah 1993; Poffenberger 1990

---

agroecological approaches to ATD merits attention, even though the organization is still in its early years. While part of its work involves the provision of support to its members, and the facilitation of mutual training and support among them, CLADES is also clearly embarked on an advocacy mission (Box

5.20). It aims to promote agroecological perspectives on development in the orthodox aid agencies, who it presumes will subsequently exert the same influence over South-based institutions receiving their funding. CLADES also engages in activities aimed at influencing Latin American bureaucracies more directly.

The strategies it pursues with this objective in mind are many. In part, it supports its members' own advocacy activities. At the same time, CLADES lobbies donors; for instance, the recent interest in the Inter-American Foundation in low input agricultural development is in part an outcome of the Foundation's contacts with CLADES.

At the same time, CLADES has set itself the task of effecting a slower, but more profound, influence on bureaucratic culture by establishing an agreement with the agronomy departments in ten different Latin American universities. Under this agreement, CLADES and its member NGOs will work with these universities to incorporate agroecological perspectives and low input production practices into the curricula of agronomy courses. Rather than retrain NARS out of certain Green Revolution ways of thinking, CLADES instead will try and influence technicians before they reach NARS and other institutions (Adriance 1992; Altieri and Yurjevic 1991).

The example of CLADES is interesting because it is an example of the creation of a southern NGO which has advocacy in both South and North as one of its main goals (it is also interesting because of the impacts it has had). Other NGOs are also active in protests directed towards the state, and in some cases towards funding agencies (Clark 1991). Particularly common themes in these lobbying actions are wider ecological issues such as the construction of dams (for example the campaign by Friends of the Earth against World Bank funding of the Sardar Sarovar project in India),[5] uncontrolled logging (e.g. the campaigns launched by Thai NGOs in the wake of 1990 floods attributable to deforestation), or the excessive use of pesticides (e.g. the campaigns co-ordinated by the International Pesticide Action Network).

In some sense, much else that NGOs engage in is a form of advocacy to the extent that their actions convey an implicit criticism of those public (and other) institutions which follow different agenda. In another sense, however, these direct forms of advocacy are qualitatively different, involving the NGO in the specific allocation of resources for lobbying. This raises the problem of representativeness much more acutely than in action oriented NGOs. While NGOs are advocating, they usually presume (or are presumed) to be advocating on behalf of the rural poor. This may be a laudable contribution, because it is often the case that the NGO has greater capacity to be an advocate than do the poor themselves (who cannot spare the time, lack the contacts, and are less able to operate in bureaucratic environments: cf. Carroll 1992). Yet on the other hand, as we have noted, most GSOs and northern NGOs are neither formally accountable to, nor composed of, the rural poor. It is therefore problematic to assume that they represent the real concerns of the poor. As we noted in Chapter 3, many NGOs are clearly motivated by some form of ideological, or professional, concern. It may well be that what they advocate is nourished more by personal politics (and ambition) than by what they have been told by the poor to be their main concern. Would an agroecological NGO lobby for cheaper agrochemicals even if it was told by farmers that their main concern was the cost of fertilisers and pesticides?

Rural social movements and democratization

Ultimately discussions of participation in ATD and empowerment of the rural poor are closely linked to the issue of rural democratization. In a recent discussion of the theme, Fox (1990a: 1) refers to rural democratization as 'an on-going process which develops, often unevenly, in the realms of both society and the state. Within civil society it involves the emergence and consolidation of social and political institutions capable of representing rural interests vis-à-vis the state'. That is, it involves a change in civil society, which becomes stronger and more assertive – and a change in government institutions, which become more responsive to the demands of that assertive civil society.

Of course, civil society is composed of a range of interest groupings, not all of them representing the concerns of the poor. In our theme, however, progressive rural democratization would involve the consolidation of groups of the rural poor and of their ability to make claims on the state. This emergence has indeed been remarked upon in the last few years' flurry of literature, particularly in Latin America, on new social movements – localized, non-class based rural popular organizations (Slater 1985; Laclau 1985).[6] Not all are rural, but some are (Harvey 1991; Bebbington et al. 1993). Such movements have been engaged in self-managed development activities as well as in questioning the state's 'top down' decision-making procedures, its failure to deliver inputs and services adequately, and the impact of development policies on indigenous cultural identities (Bebbington et al. 1993).

The role of non-membership NGOs (Carroll's GSOs) in this process of democratization is a contentious point. Radical critics of them ultimately suggest they steal the reins from these popular initiatives. There may be some truth in this, but it is arguable that GSOs still have a very important contribution to make – both to the strengthening of rural social movements, and to an enhanced responsiveness in government institutions. Let us elaborate this point.

On the one hand, while social movements represent a necessary force for rural democratization (Fox 1990a), they are not a sufficient force. This is in part because they often display poor internal democracy (despite claims to the contrary: see Carroll 1992; Landsberger and Hewitt 1970), and in part because their impact and survival depend on national policies that are often beyond their reach. As Friedmann (1992: 158) points out, '. . . to be small and local is not enough'. GSOs can, on the one hand, be a counterbalance to such antidemocratic tendencies within the organizations alongside whom they work (Carroll 1992), and can also provide valuable support during times of particularly adverse policy contexts.

At the same time, GSOs have made an important contribution to the emergence of these movements. The most obvious contribution has been the support given in their initial formation, in their legal registration, and in the on-going training in skills ranging from the technical, through the economic to the political. However, just as important are the more indirect and cumulative effects of GSO action in rural areas. These actions have, in some cases (Bebbington 1992), steadily contributed to the development of a cadre of local leaders and negotiators, to a popular awareness that different forms of grassroots development are indeed possible, and to an awareness among the rural poor that they themselves have the skills to administer such processes of change. And perhaps most critically, these GSOs have contributed to an awareness of rights of citizenship among the rural poor, an awareness that over time takes root and subsequently motivates claims they make independently of any GSO support, both on the state and often on the GSOs themselves.

In parallel with this work, GSOs in their own advocacy actions can contribute to the opening up of state structures so that they become more accountable. For example, in Bangladesh at the same time as BRAC aims to build organized landless groups for local level development action, it is also engaging with government agencies. In this latter work it provides training, technology and expertise to improve the quality of government services – for example in primary education – and responsiveness and accountability of these services to the needs of the poor.

Some analysts of NGOs' roles in mobilizing local people have reached for a people-centred terminology which sometimes verges on the mystical. None the less, such insights can constitute a much-needed antidote to the more unimaginative, functional views of NGOs. Part of the potential strength of NGOs' relationships with people is seen as their capacity for helping, as outsiders, to build 'free space' by creating an environment in which people are able to learn '. . . a new self-respect, a deeper and more assertive group identity, public skills and the values of co-operation' (Evans and Boyte 1986 quoted in Carroll 1992).[7]

The poverty reach of case-study NGOs

While firm comparative data are unavailable, the performance of the case-study NGOs in reaching the poor appears similar to that of those reviewed by Carroll (1992) and Riddell and Robinson (1993). Indeed, the criteria used by some NGOs for selecting clients specifically exclude the poorest. BAIF in India, for example, delivers artificial insemination services to small-scale owners of dairy cattle, who are likely to lie in the fifth to seventh deciles of income distribution: the poorest households do not generally own cattle. More generally, NGOs that work with technologies that imply significant risk of failing to deliver promised outputs also exclude the poorest who are the least able to countenance running such a risk.

It is also the case that the time spent in participation has a high opportunity cost to the rural poor whose main economic resource is often their time. Furthermore, the cost is probably greatest to the very poorest. They are therefore unlikely to be willing to engage in participation without the prospect of early economic gain, a factor which has contributed to the demise of a number of NGO efforts to set up local organizations. Worse still, the poorest simply may not participate in such organizations or project groups, thus reducing the poverty reach of the related NGO activity.

In the African context, there is little evidence from the case studies of specific focus on the poorest decile. Exceptions include the efforts of Langbensi agriculture station in Ghana to form women's groups (Kolbilla and Wellard 1993), and the comprehensive empowering work of ORAP in Zimbabwe, which is open to all (Ndiweni 1993; see also Box 4.6). Even here, however, those unable to provide a labour input into joint activities may be excluded from its benefits. Certainly, in other cases disquiet has been expressed over the fact that even groups recently-formed by NGOs are dominated by middle-income members: Sarch (1993) makes this point forcibly in her study of NGOs participating in the Gambian Farmer Innovation and Technology Testing Programme, and Buck's (1993) review of the Agroforestry Plots for Rural Kenya Project notes that groups formed by a local NGO, Mazingira, rarely contained representatives from the poorest households.

The poorest in South Asia may similarly be unable to allocate enough labour to productive activities, but are likely to suffer one or both of two further disadvantages: access to inadequate land and low caste status, which remains a source of systematic discrimination. Several case-study NGOs have sought to address these problems directly: PRADAN, for instance, works with low-caste groups, who are likely to be in the eighth or ninth decile, but even these have bicycles and the tools of their trade (Vasimalai 1993; see also Box 5.21). Many NGOs in Bangladesh respond to these needs by promoting income-generating activities that require little or no land, such as backyard poultry-keeping (BRAC – Mustafa *et al.* 1993; see also Box 3.9), ducks (FIVDB – Nahas 1993), and water-lifting technologies such as the landless irrigators' programmes of BRAC (Mustafa *et al.* 1993) and Proshika (Wood and Palmer-Jones 1990). Even such a low-technology device as the treadle pump (Box 5.14) with sales of 65,000/year is claimed to have created numerous casual employment opportunities for the landless. It appears clear that job creation is potentially of substantial benefit to the poorest strata, yet, outside South Asia, Carroll (1992) notes the dearth of NGOs that have made the creation of casual unskilled employment opportunities a plank of their programmes.

In the end, however, there are limits on what NGOs can say about their impact on poverty because very few actually monitor the impact of their work.

114

There are signs that this is changing (and rightly so) as donors become more demanding and want more formal monitoring and accounting systems installed. It is certainly the case that monitoring and self-evaluation have been one of NGOs' weakest points.

ISSUES ARISING FOR NGOs, GOVERNMENTS AND FUNDING AGENCIES

Who controls the agenda?

In an ideal world, local communities would determine the nature and pace of change, and external agencies would submit for their consideration a number of options consistent with expressed demands. As Korten (1990: 219) suggests: 'In authentic development an assisting agency is a participant in a development process that is community driven, community led and community owned – basic conditions for sustainability'. While the principles underlying these assertions are incontrovertible, to interpret them as implying for NGOs and GOs a purely reactive posture would excessively restrict both the scope and pace of change. In practice, external agencies may usefully be more proactive by drawing on wider spheres of knowledge in order to suggest pathways and options which have not (yet) arisen in the experiences of local groups. While perhaps fostering some form of dependence, and thus slightly diminishing empowerment in its purest interpretation, if handled sensitively these initiatives widen the range of options from which the poor can choose, and so have the potential to contribute to acceptable change.

Interaction of this kind is far removed from the tendency by many development agencies (governmental, externally-funded projects and, to some extent, NGOs) to seek 'participation' simply in order to get local agreement to a predetermined agenda. Superficial interaction of this kind can have various effects. It may mean that external agencies discuss development proposals only with a few unrepresentative community leaders; it may lead to situations where community meetings, sensing what type of external assistance is in the offing, tell visiting experts 'what they want to hear'; or it may mean that opportunities are opened up for powerful groups to take control of the agenda. The term 'participation' is increasingly being abused in this way. An example exhibiting some of these shortcomings is provided by the Philippine Ecosystem Research and Development Bureau, a quasi-autonomous agency which, when under contract to the Department of Environment and Natural Resources, worked with (and, in some cases, created) local groups to take up largely pre-determined agri-livestock technologies (Box 4.9).

For NGOs committed to the empowerment of local groups, the question of who controls the agenda has particular significance. The logic behind a commitment to empowerment is that the NGO should be willing to share with local groups not only the management of a project or programme, but also the

115

wider questions of the NGO's overall agenda and in particular whether it should continue to be present in a particular area. Some would suggest that non-membership NGOs should, logically, go further and allow the local organizations which they service to elect representatives to the NGO's board. In this way they might emulate some of the formal procedures for representation adopted by MSOs, although these, as Carroll (1992) and Fox (1990) warn, have in many cases proved to be democratic more in concept than in practice. Nevertheless, by failing to democratize links between themselves and local membership organizations, it is clear that NGOs lay themselves open to accusations of 'non-representativeness' and 'lack of accountability' by governments, and provide an easy excuse for governments to ignore those NGOs whose views challenge their own.

Timing the withdrawal of NGO support from local membership organizations

As we noted, many NGOs pursuing empowering approaches see their interaction with local membership organizations as temporary: they aim to strengthen such organizations to the point at which they are able to take on many of the functions performed initially by the NGO. However, successful withdrawal (as opposed to project termination) is easier to achieve in rhetoric than in practice. Indeed, very few NGO projects from those documented across the three regions appear to have reached this stage. The contrasting experiences of two of these are informative.

Box 4.9 Agri-livestock technologies in the Philippine uplands – participation or predetermined agenda?

In the Philippines, the Government's Ecosystems Research and Development Bureau (ERDB) of the Department of the Environment and Natural Resources (DENR) is working with local people's organizations (POs) in order to create a more effective framework for the delivery of agri-livestock inputs to upland areas where there was little government presence during the Marcos years and the reputation of government extension services was very poor. In communities where POs do not already exist or are considered unsuitable for becoming the recipients of government supplied inputs, the government is helping to 'form' the necessary organizational structures. Although this approach gives the government better access to communities and their perceived needs and can therefore provide a better standard of services, the situation in some localities remains a long way from the levels of participation implied by government rhetoric, and amounts to little more than participation in a development agenda which the DENR has determined internally.

Source: Tomboc and Reyes 1993

PRADAN's work in promoting chrome-leather tanning (Vasimalai 1993; see also Box 5.21) focused on close interaction with a single flayers' co-operative in facilitating their adoption of a single technology. Within three years, they were sufficiently in command of the technology, and well enough linked with government support services and with the private commercial sector for PRADAN to phase out its support.

The experience of Catholic Relief Services (CRS) in The Gambia (Owens 1993) provides a marked contrast: its agricultural activities embraced a range of crops, including sesame, fruits and vegetables. CRS was working with 50,000 women in sesame growers' associations throughout the country at the crop's peak level of production, and with numerous other local organizations in further crops. For sesame alone, research at CRS's central research station and at two other sites in different agro-ecological zones contributed to (but were by no means the only factors in) a spectacular increase in cropped area from 30 ha in 1983 to 12,000 ha in 1986. Despite some CRS success in facilitating the introduction of processing equipment, problems with marketing were a principal factor in the crop's decline to some 2,000 ha in 1990.

The crop's decline was one factor which prompted an external review to recommend a shift away from on-station research to farmer-managed trials. This was implemented in 1989, but the inadequate preparation of CRS's Village Monitors for this work meant that fewer than twenty of the 124 micro-plots that had been established were actually harvested and discussed with farmers.

Further major changes in 1991 were grounded in the belief that local communities had become capable of conducting and interpreting their own trials. CRS's own research effort was scaled down, and the team of fourteen Village Monitors disbanded. This community-based experimentation and extension approach relied on the recruitment of new facilitators who would live at village level and promote both farmers' experimentation and farmer-to-farmer dissemination.

While firm conclusions cannot yet be reached, the wisdom of recent changes at CRS must be queried in view of the facts that:

- The 1989 experience with farmer experimentation produced poor results;
- The Department of Agricultural Research is too weak to meet research requirements that otherwise might have been met by CRS's own research effort (see Sarch 1993; Sarch and Copestake 1993);
- The local groups which CRS had formed were in 1992 being encouraged to conduct 'needs assessment' using participatory rural appraisal techniques with which they had been unfamiliar. Early reports (Gilbert, pers. comm.) suggest that they are doing so not because they see a need for them, but because of a CRS policy that such techniques should be used by local groups.

While it would be unreasonable to draw starkly contrasting conclusions from these two experiences, they suggest that it is easier for an NGO to withdraw successfully from situations in which highly participatory work with a single group has led to the adoption of a single new technology. Conversely, it is much more difficult to design withdrawal paths from situations in which numerous technologies have been developed through more tenuous links with a large number of less cohesive local groups. Similarly, in these more complex situations, there is great risk of being too rapid in scaling down the levels of NGO service provided hitherto – with the result that earlier gains in the project might be lost, by passing on too many tasks to local farmers too quickly.

Is the scope for less-than-ideal improvements being under-estimated?

Numerous observers and practitioners have suggested that participation focusing narrowly on development technology is an inadequate basis for sustainable development. F.H. Abed, for instance, as chairman of the NGO Bangladesh Rural Advancement Committee draws on concepts developed in Brazil by Paolo Freire (1972) when he notes that '. . . [the] adoption of innovations, which is the goal of any extension programme, cannot take place unless the capacity of the receivers is properly developed' (Abed 1991: 4).

While laudable as a general principle, in practice this notion generates severe tensions for agencies operating at sectoral levels, which is the typical mode of operation of government departments. Government research and extension services, for instance, have mandates limited to agricultural technology (and, increasingly are provided with budgets which restrict them to only a part of that mandate). While they may, with considerable time and effort, achieve with limited numbers of clients the collaborative or collegiate levels of participation indicated by Biggs (1989), their mandate does not allow them to extend their focus beyond agricultural technology, so that the types of interaction summarized in the lower half of Figure 4.1 are unlikely to be achievable.

Early populist critiques of agricultural research seeking deeper levels of participation demonstrated an impatience with the slow pace of change in government service provision that bordered on dismissiveness.[8] In our view, however, the very substantial scope for improving client-orientation in government services, together with their wider coverage than those of NGOs, means that significant benefits to clients can be achieved by fuller exploitation of the approaches (diagnosis, experimentation, feedback) characteristic of NARS. This is not to deny that the introduction of comprehensive institutional and methodological change, such as that advocated by Pretty and Chambers (1992) would lead to deeper participation and greatly enhanced benefits to clients; it merely suggests that the less spectacular but none the less very substantial improvements that can be made while we are awaiting the revolution should not be ignored. One of the best examples of how the approaches currently available to government can be more fully utilized has, in fact, been provided by an NGO: BAIF (Box 4.10) in India has demonstrated unequivocally how an 'issue' focus and careful monitoring and feedback can lead to wide technology adoption.

Box 4.10 NGO success without participation – the Bharatiya Agro-Industries Foundation (BAIF) in India

---

Established in 1967 to develop new dairy cattle technologies, initially for workers in the sugar cane industry, BAIF is an unusual NGO in many respects:

- Its research farm employs over 40 professional scientists, several of whom formerly worked in the public sector, and has pioneered frozen semen artificial insemination technology in India;
- It operates its own village-level AI service, which has reached 1.5 million families in six states and is responsible for almost 10 per cent of cross-bred dairy cattle in India;
- The same service provides advice and technologies for improved animal nutrition deriving from the same research farm, and for animal health developed by a subsidiary which researches into and produces vaccines;
- BAIF has attracted funding from foreign donors, from Indian government research and development budgets and from the private commercial sector (under concessional tax provisions).

BAIF has grown into a highly successful organization: its research commands respect among peers: it provides services which farmers are willing to pay for, and it does so more cost-effectively than public sector organizations. Although its technologies are geared to the poor rather than to the poorest, there can be little doubt that the increased incomes and employment opportunities that they have created have benefited the poorest. Yet its philosophy is to deal with farmers on an individual, not a group, basis, and its focus is narrowly on *economic* improvement. Concepts of participation and empowerment as they are conventionally articulated have little influence on its strategies or activities. Few would suggest that the decision to focus on AI in the first place, or to introduce it in specific locations, is based on formal 'needs surveys'. These decisions draw at least as heavily on well-informed scientific opinion as on rural voices.

What BAIF *has* demonstrated are the gains that can be made through issue-focused research and through careful monitoring of the main technology that it has offered – both to check its performance and to identify additional measures (health, nutrition) needed to supplement it. It is ironic that, as an NGO, BAIF has restricted itself largely to the methods and approaches which are nominally in the government's preserve, and yet has demonstrated the gains that these can offer if adequately focused and fully exploited.

Source: Satish and Farrington 1990; Bhat and Satish 1993

Finally, it must also be recognized that NGOs incur substantial resource costs when pursuing deeper levels of participation, costs that would be more difficult for government to absorb, and that ultimately might be difficult to justify in general. PRADAN, for instance, took six months of regular contacts in order to gain the confidence of a flayers' co-operative, and a further two and a half years in supporting the construction and management of a new tanning plant (Box 5.21).

These observations lead us to a number of conclusions. First, with persistence, patience, and considerable resources 'ideal' concepts of participation can be achieved by NGOs in highly localized conditions. Second, the sectoral mandate of government agencies is likely to prevent them from achieving much depth of participation but this does not prevent them from developing, for instance, agricultural technologies and management practices which meet at least some of the needs of the poor. Third, if sensitively designed, such changes can also be of benefit to the poorest through the creation of opportunities for casual, unskilled employment. Given the resource constraints currently facing GOs, to pursue the minimum levels of participation necessary to discover what farmers want is an understandable strategy which may not be far from optimal for those tied to a narrow sectoral focus.

In principle, and in a less functionalist view, close integration among government line departments, among NGOs and among local government development efforts, and among all three categories, could offer prospects for more profound levels of participation over wide areas. But, as suggested by evidence discussed in the following chapter, this has often proved difficult to achieve. Indeed, the costs of participation have been one reason for this limited progress. One organization's participation in the activities of another implies time, staff and other resource costs. Many organizations, seeing few returns to such participation, decide ultimately to go it alone.

CONCLUSIONS

This chapter has focused on NGOs' interactions with the rural poor. It began by attempting to identify where poverty occurs, and what future trends can be expected: while around two-thirds of those in poverty are currently found in Asia, absolute numbers there are likely to decline slightly over the next decade, whereas they will increase substantially in Africa. An understanding of the *processes* underlying poverty is essential if appropriate interventions are to be designed.

NGOs' ability to alleviate poverty and to work with poor people has been the focus of much uncritical speculation about their potential – not least from NGOs themselves. Two recent empirical studies reviewed in this chapter provide some support for these claims, but also draw attention to the fact that very few NGO income-generating projects manage to reach the poorest, that a substantial proportion of projects are financially non-viable, that, outside S. Asia, few NGOs have worked to create new employment opportunities despite the potential poverty alleviating impact of such a strategy, and that much of NGOs' work is conducted in isolation from the wider policy arena. Consequently, many NGOs do in effect (if not by intention) end up working primarily as service deliverers, rather than as sustainable poverty alleviators or as policy innovators. To that extent, we suggested that the instrumentalist perceptions of NGOs held by many donors and governments, who see NGOs as means of getting services out to the poor, might in fact be a more accurate interpretation of what NGOs have in fact done than are the interpretations that NGOs have of their own work.

We then looked at the empirical material on NGOs' work with the rural poor in agricultural technology development and dissemination assembled for this study. Likewise, this material, while showing some quantitative outputs, is most significantly based on the qualitative processes that NGOs initiate (participation, empowerment). A simple conceptual framework is presented and illustrated by examples to demonstrate how their work can usefully be analysed both in terms of the *depth* of participation achieved, and the *scope* of subject matter focus. Efforts towards deeply participatory ('empowering') approaches made by many NGOs inevitably lead them also into a wide thematic focus, since the livelihood constraints identified through processes of conscientization are rarely sector specific. This can lead to overstretching and to disillusion among clients since NGOs' technical and organizational ability to facilitate improvement in a wide range of issues is inevitably limited. GOs, by contrast, generally have a narrow sectoral mandate, which constrains their ability to enter the types of deeper dialogue with clients that are likely to bring onto the agenda cross-sectoral issues. Nevertheless, within the narrow context of identifying, developing and delivering adoptable technologies, there remains scope for GOs' performance to be enhanced – and their poverty reach to be improved – by better use of the tools currently at their disposal, viz. the conventional techniques of project preparation, monitoring and feedback, supplemented by whatever slight degrees of participation they can achieve. While we believe that NGOs' concern with more profound levels of participation and empowerment is important (we too are impressed by NGOs), we demonstrate in Chapter 5 that fortunately, a number of them are concerned to identify how less radical change in the government sector might be fostered. We argue that these latter strategies may generate positive outcomes for the rural poor.

NOTES

1 Mellor and Chambers would, however, disagree on many other points – not least whether development should concentrate attention on these low potential areas (Chambers) or essentially write them off and focus on the higher potential areas (Mellor).

2 Concurrent with, but independent from, the Riddell and Robinson study, a group of Dutch researchers evaluated the performance of a similar-sized sample of Dutch NGO sponsored projects, and came to very similar conclusions (Wils 1990).

3 For five projects benefits clearly outweighed costs; the objectives of a further five

were achieved, but at high cost; the objectives of two projects were not achieved, and, in a further four, no clear assessment of cost-effectiveness could be made.

4 The work of Mag-uugmad Foundation Inc. in the Philippines helps, at farmers' request, to incorporate soil and water conservation into their traditional labour-sharing arrangements, and, at the same time presses for secure access to forest land. It thus approaches agricultural issues from within the context of wider social change.

5 Open letter from Friends of the Earth to Baroness Chalker of the UK Overseas Development Administration on 13 October 1992.

6 Similarly, in many Asian countries, the popular and NGO sectors became an arena for rural activists after overt political activity became the object of state repression: for example, the Philippines under Marcos, Bangladesh in the 1970s and present day Indonesia.

7 This is part of what Hirschmann (1984; also Uphoff 1993: 28) call 'social energy' – a creative, transformative phenomenon which committed NGO organizers can some-times bring into local situations. Uphoff stresses the ability of NGOs to affirm the value of such phenomena as ideas, ideals and friendship which are generally undervalued in economistic approaches to development and draw on them as positive catalysts for change.

8 See for instance, Chambers and Jiggins (1986).

5

RELUCTANT PARTNERSHIPS, HOSTILE CONFRONTATIONS, OR PRODUCTIVE SYNERGIES? POSSIBILITIES FOR NGO–STATE INTERACTIONS

[NGOs can] . . . oppose the State, complement it, or reform it, but they cannot ignore it.

(Clark 1991: 75)

'Should they dance or shouldn't they?'[1] So far in this book we have tried to have it both ways. We have pointed in the first two chapters to all the very understandable reasons NGOs might give for not working with state institutions. Democracies are fragile, states are inefficient, armies are still twitching in the wings, and NGOs are in danger of being co-opted as instruments of government policy. Yet, lest we appear uncritical champions of the NGO cause, in the third and fourth chapters we have pointed to a range of weaknesses that in some sense are inherent to NGOs. Aside from the idealistic overstatements of their strengths,[2] there are structural limits on what NGOs can achieve when acting alone: limits on their technological capabilities, their impacts on poverty, and their capacity to be legitimate, and accurate, representatives of the concerns of the rural poor. Not all these limits would be addressed through engaging with state institutions, but some might – in particular, access to technological resources, and to the policy making process. At the same time, we have suggested that problems of co-ordination of NGO actions have sometimes been very severe and that in some circumstances state led co-ordination mechanisms might be a valid means of improving synergy among NGOs and between them and the state's own agencies.

The decision to dance, or not, is ultimately a country-specific one. We drew attention in Chapter 2 to some of the variability that there is among different countries as regards the amenability (or otherwise) of the overall policy and political environment for fruitful NGO–state relations – in some circumstances it would be very difficult to justify any attempt by an NGO to collaborate with the state. However, we have also suggested that in many places the structural and funding forces seem such that NGOs and governments will necessarily have to consider working with each other.

In the end, however, there seem to us some very good reasons for NGOs accepting, reluctantly or not, the invitation of a dance. We begin this chapter with a summary of those reasons. If the dance is to go ahead, then we need to know something about the style and success of the steps that have been tried out before, in order to add strength to the argument, to learn about the range of possible dances on offer, and to learn from past experiences where partners have stumbled so as to get the steps right in the future. We also need to draw on insights provided by interaction between NGOs and their other partners: we examine particularly the contribution that linkages with the private commercial sector can provide to scaling up, to income generation and to the sustainability of grassroots organizations. This discussion occupies the bulk of the chapter, and is based on the case study material. This is analysed in six sections:

- The first examines some of the implications for NGOs of political, economic and administrative changes being introduced by government;
- The second focuses specifically on the interacting roles that NGOs and NARS have played in the agricultural technology system and identifies linkage mechanisms that have helped to make those interactions more effective;
- The third examines linkages established as a result of government initiatives;
- The fourth examines a wide range of *interactions*, many of them still in the formative stage, encompassing initiatives taken by NGOs which they intend for eventual scaling up by government, and NGOs' interaction with a range of government agencies as trainers and catalysts;
- The fifth examines the potential NGOs have for combining conflictive and collaborative interaction;
- The sixth reviews links established between NGOs and the private commercial sector.

A NEW DANCE FLOOR?

The combined effect of structural adjustment and democratization is increasingly creating an environment which, we argue, favours closer collaboration between NGOs and the state. In doing so, it also makes the limitations of NGOs working alone seem that much more serious, and a go-it-alone strategy that much more unjustifiable.

Political economic changes

As we noted in the first chapter, the 1980s saw, albeit to different extents in different countries, a series of political, institutional and economic changes that have created an environment in which there is more scope for, and indeed pressure for, closer co-operation between NGOs and the state (Bebbington and Thiele 1993). On the one hand, the trend towards political democratization,

124

and the consequent weakening of authoritarian tendencies, has made it more conceivable for NGOs to work with governments, and has in some cases made government more flexible. To the extent that donors are setting more efficient and accountable government (good government) as a condition of financial aid, these positive and enabling tendencies are given greater momentum. At the same time, administrative decentralization in the public sector has made government more accessible to many NGOs. While decentralization has associated risks (that local government will be given responsibilities but no funds, or that local governments will be captured by local elites), most NGOs in this study have seen it as a positive step forward.

In a perverse way structural adjustment programmes have also created more scope for closer co-operation, because by reducing the state's administrative capacity they have forced the state to search out for new 'partners' in the private voluntary and commercial sectors. Of course, the perplexing question for NGOs is whether, if they accept any of the opportunities for collaboration that are in some sense related to public sector retrenchments, they are implicitly endorsing policies whose cumulative effect is to question the principle that the state has a responsibility to provide welfare to its citizens (Bebbington and Thiele 1993). On the other hand, if they turn such opportunities down, they are losing a chance to exert at least some leverage over government policy and actions.

Reforming the NARS – creating space for NGOs

These wider political economic processes have tended to create a more favourable general environment for fruitful NGO–state interactions. How far they influence the NARS varies among countries. In Latin America, it is clear that the climate of public sector withdrawal from service provision has influenced recent trends in the NARS. In some countries, the NARS is withdrawing from extension services, and expects other institutions (including NGOs) to provide and finance these (e.g. Bolivia and Peru). For instance, the Bolivian NARS is divided into two parts: CIAT serving the lowland department of Santa Cruz, and IBTA (the Institute for Agricultural Technology) serving the rest of the country. Both used to have extension services. In the mid-1980s, however, as a result of funding cutbacks, CIAT lost its extension services, and has since then established working links with the NGOs in Santa Cruz which now disseminate CIAT technologies. Similar funding crises in IBTA led it also to decide in 1990 to relinquish extension activities and seek out working links with NGOs doing extension (though it has not yet been as successful as CIAT in this – Bebbington and Thiele 1993: Ch. 5).

In other cases, the state retains a certain financial obligation for extension, but increasingly wants this work to be done by private agencies, including NGOs (e.g. Chile, Colombia). For instance, in Chile, the public sector ceased to implement extension activities in the early 1980s, and contracted these out to the private sector. Since the end of the Pinochet dictatorship, NGOs have been able to win these contracts. Similarly, the new government has brought NGOs into the management of its on-farm and regional research programme. In these restructurings, it is generally the case that the NARS identifies a new role for itself as a service to these so-called 'intermediate users' of its technologies – rather than as a direct service to farmers. They thus begin to make themselves, and their resources, more available and open to NGOs. In conjunction with these changes is an increasing interest on the part of the NARS to co-fund activities with NGOs and other organizations.

Once again NGOs who disagree with such state withdrawal are faced with the question as to whether, if they engage in these new relationships, they are effectively (if not intentionally) endorsing a policy with which they disagree. On the other hand these changes do (in cases where the NARS is not so chronically weak and underfunded that it has little to offer anybody) make it easier for the NGOs to use the NARS as a technical support service in a way that has been more difficult in the past. At the same time, the NARS are coming to the NGOs considerably weakened. As such, the NGOs have a certain bargaining power, and have the potential to use it to gain a certain influence over the agenda of the NARS, to orient them more towards the needs of the rural poor.

These changes are particularly marked in Latin America. In Africa, withdrawal is also occurring, but more as part of the generalized crisis, and in places collapse, of the NARS. In Asia, NARS remain more centralized. None the less, as we will see, changes leading the NARS to show greater interest in opening up to NGOs are apparent. In the end, whether this is really a new, more attractive and enabling dance floor or not, depends on the country. In general though, it seems that barriers to fruitful interaction are not as severe as they were.

Together these changes have had the effect of making government more accessible, and weakening its hand to undermine NGOs and other civic associations with which it has differences of political opinion. Furthermore, these changes have reduced the capacity of the NARS to have any great impact on rural poverty (and growth) if they act alone. At the same time, as subsidies on inputs are removed and NARS are faced with the increased necessity to find low external input and low cost production technologies, they have found their cupboards rather bare – most of them have little that is agroecological with which to face this new environment.

On the other hand, some of these changes have also made some of the limitations and failings of NGOs that much more apparent. The fact that many governments are now in some sense electorally accountable may not mean that they are perfectly democratic regimes, but it does at least highlight the fact that the government is now popularly elected whereas NGOs are not. Similarly, in a perverse way, the very negative impacts of adjustment policies on the poor and on many NGO programmes has highlighted the policy dependence and ultimate vulnerability of any impact NGOs may have on rural welfare. And finally, the increasing amount of funds being channelled through NGOs is leading to an institutional proliferation so out of control that problems of NGO co-ordination and quality control are now more severe than they have ever been in the past.

For these reasons, and those elaborated in earlier chapters, it seems time to assess how far closer NGO–state relationships might be an adequate response to this new institutional and political economic landscape.

LINKAGES BETWEEN NGOs AND NARS – A CONCEPTUAL NOTE[3]

Consideration of any future strategy of forging closer links between NGOs and NARS in ATD must be based on an assessment of experience to date. Though our sample of case study NGOs and NARS was biased by the deliberate attempt to document cases where there had been inter-institutional interaction, the range and extent of interaction that was uncovered was striking, given NGOs' rhetoric of maintaining a distance from state institutions.

This interaction occurred at two main levels. On the one hand, there were interactions which stemmed from the fact that different institutions played roles in different parts of the ATS, and so were able to draw on the work of the other in order to enhance the impact of their own activities. In these cases, one institution might, for instance, have drawn on technologies developed by the other, though without any formal contact (or at times without any contact at all) – for example, when NGOs worked with technologies developed by government research institutes, but never in fact talked to those institutes. On the other hand, and somewhat different from the first level of interaction, were cases where the relationship between both sides was recognized by the actors, and involved *conscious co-ordination and contacts* – in other words, an interaction in which some form of *mechanism* was used for managing the relationship between the two.

In what follows we begin with a general discussion of the interactions (managed or *de facto*) emerging from the inter-institutional division of roles, or tasks, in the ATS.[4] We then focus on the cases where these relationships were managed, concentrating on the mechanisms through which the relationships were structured, developed and sustained. Mechanisms may be *structural* – that is, permitting influence by one side on the institutional or organizational characteristics of the other – or *operational* – that is, activity- and project-specific. They may also be formal or informal.

As we talk about these different forms of interaction it is helpful to disaggregate them, (see Figure 5.1). *Interaction* is itself the broadest term. By interactions we refer to situations where the actions of one institution are influenced by, dependent on, or oriented toward the actions of another institution. However, the quality of this interaction, whether it is collaborative

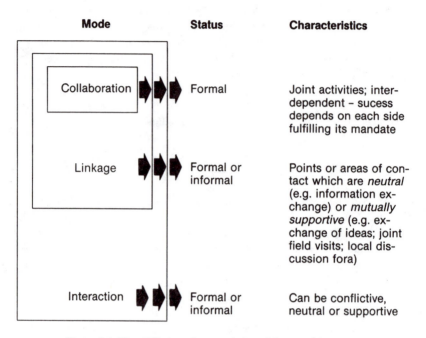

Figure 5.1 The differing characteristics of forms of interaction

or conflictive, managed or casual, remains unspecified. As a sub-set of interactions, *collaboration* implies a formalized dependence of one partner on another for at least part of the success of its activities, as when, for instance, a GO might contract NGOs to disseminate technologies that it has developed. As another sub-set, *linkages* occupy an intermediate position, implying positive interaction that may be formal or informal, but is of a less mutually dependent kind than that designated by the term *collaboration*.

We have talked in earlier chapters about more conflictive interactions between NGOs and the state. This chapter thus concentrates more on the collaborative actions.

TECHNOLOGY PATHS AND THE DIVISION OF TASKS AMONG NARS AND NGOs IN THE ATS

NGOs and NARS could play one or more of a wide range of roles in interacting with each other in ATD. Thiele (Chapter 6 in Bebbington and Thiele 1993) has developed the concept of technology 'paths' as an aid to understanding the roles that might be played. Following Biggs (1989) he suggests that a multiplicity of institutions (public sector, NGO, farmers, and private commercial sector, whether local, national or international) might each be involved in one or more stages of the development of a technology as it passes along a path from basic research, through strategic, applied and adaptive research to dissemination (see Chapter 2).[5] In order to keep the number of possible institutional roles to a manageable size, Thiele restricts his analysis to only two types of institution – NGOs and NARS – at the three stages of applied and adaptive research and dissemination. This yields eight sets of roles.

The analysis presented here is a further simplification of Thiele's: it combines adaptive and applied research into a single category, since in practice NGOs very rarely engage in applied research,[6] and eliminates the most orthodox category in which NARS occupy all three stages to the exclusion of NGOs.[7] The resulting four sets of roles are presented at the heads of columns in Table 5.1, and comprise:

1 NGOs disseminate the results of research conducted by NARS;
2 NGOs and NARS conduct research jointly, and NGOs disseminate;
3 NARS disseminate the results of research conducted by NGOs;
4 NGOs disseminate the results of their own research.

It is worth analysing in more detail the four different combinations of roles set out in Table 5.1 and the contexts from which they originate.

NGOs disseminate the results of research conducted by NARS

The first technology path, in which NGOs disseminate technologies developed by NARS without adapting them, occurred in four instances each in Asia and S. America, and two in Africa.[8] The low representation of cases in this category may in part be explained by the fact that little so far has come out of the NARS to meet many NGOs' concern with low input agriculture, agroforestry, and soil and water conservation. Furthermore, most of these NGOs who initially disseminated NARS technology uncritically, subsequently began their own adaptive research once they found the technologies to be unsuitable for their clients (see AGRARIA, Box 5.2). In this sense, those NGOs we place in this first column, also appear in the second, and the transition from one form of interaction to another may be a characteristic experience for many NGOs.

NGOs and NARS conduct research jointly

In the second technology path, in which NGOs and NARS conduct research jointly, the NGO generally operates in a more obviously, and often on-farm, 'adaptive' mode than the NARS (the second column of Table 5.1). This is the single most important path in all three regions. Two-thirds of the Latin America case studies, half of those in Africa and one-third of those in Asia fall under this heading. Its pre-eminence is partly attributable to NGOs' frequent practice of adapting and 'scaling down' technologies originally developed by NARS for resource-rich farmers.

Table 5.1 Roles played by NARS and NGOs in the development of technology and of methodologies

NGO	NARS research, NGOs disseminate	NGOs and NARS conduct research jointly, NGOs disseminate	NGOs research, NARS disseminate	NGOs research and disseminate
AFRICA				
KEFRI/CARE/Mazingira Machakos NGOs	Soil and water conservation technology	Agroforestry	Dissemination methods	
ENDA GVAM		Maize, millet, sorghum		Small grains[1] Vegetables
ACDEP/Langbensi/TAAP		Varietal/management trials	Participatory research methods Animal traction	
FITT		Varietal/management trials and dissemination		
CRS				Sesame production/ processing Animal traction
Gambian NGOs Rodale/CRAR ARAF/ISRA		Trials on organic matter[2] Composting; application of organic matter		
ASIA				
CARE FIVDB				Vegetables Ducks[3]
BRAC	Poultry[1]			
Proshika	Livestock[5]			
MCC		Fruitfly trapping; vegetables		
MCC			Soya[6]	
PRADAN	Raw silk Mushrooms Hides and skins			
AWS		Caterpillar on castor Chillies		
RKM				
AKRSP			FSR[7]	
BAIF		Animal disease monitoring[8] Forage crops[8]	Training methods	Artificial insemination Vaccine production

IIRR	Integrated livestock			
ERDB	Rice-fish culture			
FSRI/CUSO/ATA				
ATA			Dissemination methods	Organic vegetables
LATIN AMERICA				
CESA Ec	Potato	Potato		
CAAP				Andean crops
FUNDAEC			Cropping system	
SEPAS		Blackberry, irrigation		
IDEAS		Cocoa		
CIED		Rice		Agroecology
CEIBO	Maize			Agroecology
CIPCA		Agroforestry		Cocoa
CESA Bol		Protected horticultural environments		
Kohl (Bolivian NGOs)				
GIA	Beans and maize system, Pastures, Wheat		Beans and maize system	
AGRARIA	Beans and maize system, Pastures, Wheat			

Notes:

1 Dissemination of the hybrid maize required considerable innovation by the NGO in the organization of credit and input supply.

2 An important facet of Rodale's work is that it sees itself as a bridge between the NARS (ISRA) and other NGOs, inviting the latter to participate in trials which it has jointly conceived with ISRA.

3 Some small-scale, informal links with Bangladesh Agriculture University.

4 BRAC's programmes disseminate technologies (vaccines; improved breeding stock) made available through government monopoly provision, but which have been the subject of very little GO research (in some cases, none). By contrast, BRAC has developed robust social organizational models capable of sustaining implementation of these technologies by the rural poor. Some training of paraprofessional poultry workers was undertaken jointly by BRAC and the Department of Livestock.

5 Similar observations apply as made about BRAC in note 4.

6 Following the collapse of earlier joint research, MCC persevered and produced varieties which the extension services are now disseminating.

7 Some technologies (e.g. rice varieties) tested by RKM are being disseminated by the public sector, but this set of relationships is particularly complex, involving ICAR research institutes and state agricultural universities in an FSR network, to which all have contributed.

8 BAIF was involved in this research at the same level as GOs, since it was part of a research/monitoring network (All-India Coordinated Research Programme).

Examples of why and how NGOs have disseminated, and in some cases adapted technologies produced by NARS for large farms to small farmers' needs are presented in Boxes 5.1–5.3. Box 5.1 provides an example where an NGO provided the means (primarily credit) to facilitate the dissemination of NARS technologies to small farmers who previously lacked access to them; Boxes 5.2 and 5.3 present cases of NGOs who have also adapted that technology to their clients' needs.

Box 5.1 Silveira House in Zimbabwe: a response to smallholders' demands for hybrid maize 1968–83

Silveira House, an NGO based in the Catholic Church, was one of the few NGOs to work with small farmers prior to independence, at a time when hybrid maize seed had been developed for commercial farmers. Small farmers saw the technology's potential, and asked Silveira House to help with supplies of seed, credit and fertilizer. In response, Silveira House set up a 'pump-priming' group-based revolving loan scheme and assisted with sales through the Grain Marketing Board, to which smallholders did not have individual access. Loans under the scheme were provided interest free for a maximum of three years to any one person. Silveira House also provided demonstrations and training (in which government extension staff participated), and sponsored competitions and field days. By 1979–80, 274 groups had been formed, comprising 2895 members; their maize yields had increased approximately fivefold, and by 1981 the groups purchased over Z$5 million worth of fertilizer. As the Agriculture Finance Corporation made credit more freely available to farmers after independence, the numbers taking loans in Silveira House's areas of operation rose fourfold, but since the schemes were not operated with the same levels of discipline or peer pressure, repayment rates fell from 95 per cent to 60 per cent in 1980–81, and to 71 per cent in 1981–2.

Source: McGarry, 1993

NARS disseminate the results of research conducted by NGOs

The cases listed in the third column of Table 5.1 represent technology paths in which the NGO carries out applied or adaptive research for subsequent dissemination by the NARS. Few examples of this type of linkage are found among the case studies, at least in part because information flows about research results from NGOs to NARS are even weaker than flows from NARS to NGOs. The only examples among the Latin America case studies are those of GIA (Sotomayor 1991; see also Bebbington and Thiele 1993) and FUNDAEC (Arbab 1988). The latter has carried out research into alternative production systems for *campesinos* since 1973, in the Cauca Department of Colombia, identifying and evaluating ten different production sub-systems on farmers' fields. At present FUNDAEC is seeking to pass on its research results

---

Box 5.2 AGRARIA, Chile: from selection to adjustment of technology

---

In 1991, the Chilean NGO, AGRARIA managed nine agricultural development projects, in addition to seven projects on contract to the government institute responsible for extension, INDAP. One of its independent projects is in Sauzal. AGRARIA's experience in Sauzal shows how the NGO has continually adapted technology originating from the NARS.

In 1984 AGRARIA was conducting research independently of the NARS. It pursued commodity-based research aimed at maximizing yields of wheat, lentils, oats, chickpeas and sorghum. A complex multiple factorial design was used for experiments, in an attempt to reproduce the techniques of an experimental station. This approach was adopted because of the total absence of relations with the National Institute for Agricultural Research (INIA) (a fruit of mutual mistrust under the Pinochet dictatorship), and the orientation of the NARS towards large scale farming. AGRARIA soon concluded that its limited resources, the lack of training of its technical teams and the availability of information about yield maximizing technologies limited the usefulness of this type of research.

In 1986 AGRARIA changed its strategy to one of adjusting technologies already developed in INIA's experimental stations. Instead of seeking yield maximising technologies, the NGO began to seek the technologies that would perform best within the *campesino* context. For example, research on wheat used much lower doses of fertilizer which were more appropriate to small farmer conditions.

In 1988 AGRARIA, continuing the strategy based on technology adjustment, introduced a systems focus. An informal research proposal was drawn up, together with specialists from INIA. In this stage the now informal relationship between certain scientists at INIA and the Sauzal project was one of reciprocal support. A relatively continuous flow of technical information from the experimental station to AGRARIA was established. AGRARIA, in turn, through its experimental work under farmer conditions, passed results to INIA's specialists. AGRARIA contended that it was necessary to increase the productivity of systems rather than work with isolated components. It succeeded in demonstrating to INIA researchers the viability of a system based on wheat and a naturally occurring legume forage *hualputra*.

Source: Aguirre and Namdar-Irani 1991

---

to public sector institutions for wider dissemination. As a first step, FUNDAEC organized courses for public sector extension staff, to familiarize them with its research results.[9]

In Asia, the work by MCC on soya in Bangladesh (Box 5.4) provides a remarkable example of perseverance in research by an NGO after the NARS had given up, which then allowed the national extension services to capitalize on the findings of the NGO's research. Other examples of shared technology development and dissemination in these paths include rice varieties identified under Ramakrishna Mission's (RKM's) FSR programme in eastern India, which were subsequently incorporated into the government's demonstration

---

Box 5.3 Adaptive research and dissemination of rice-fish culture in N.E. Thailand

---

Although formerly widespread in Central Thailand, rice-fish culture was progressively replaced by more profitable rice technologies from the early 1970s which relied on high inputs of agrochemicals incompatible with fish production. At the same time, rice-fish culture began to spread to the less intensively farmed areas of N.E. Thailand, following rapid declines in traditional wild fish sources.

NGOs were among the first to notice the increasing popularity of rice-fish farming in N.E. Thailand, and to bring it to the attention of several government agencies. CUSO, for example, helped to convene a workshop in 1983 to bring together the experience of several GOs and NGOs in this area. At the same time, conventional patterns of on-station agricultural research followed by researcher-managed on-farm trials had proved slow to produce results and were being replaced as a newly-established Farming Systems Research Institute (FSRI) produced responses to farmers' problems through interdisciplinary farmer-managed research. While the Fisheries Department's advice was incorporated informally into this work, fisheries professionals were reluctant to join the staff of the FSRI for want of a career structure. CUSO provided four volunteer fisheries scientists to the FSRI, thereby enhancing its appreciation of aquaculture, strengthening the implementation of rice-fish studies, and paving the way for closer FSRI/DoF links.

Experimentation by the FSRI with various pond-dyke configurations and management systems was observed by a number of NGOs, one of which (the Appropriate Technology Association (ATA)), adapted some of the FSRI systems into its own work to develop rice-fish farming systems integrated with vegetable, fruit tree and livestock production. Adoption of these technologies spread rapidly, partly as a result of ATA's work in training and in developing farmer-to-farmer dissemination methods.

Source: Jonjuabsong and Hwai-Kham 1993

---

programme (Chakraborty *et al.* 1993), and the animal traction technologies developed by inter-linked NGOs in northern Ghana (ACDEP/Langbensi/ TAAP) which were taken up by the German-Ghanaian Agricultural Development Project (Alebikiya 1993; Kolbilla and Wellard 1993; Millar 1991).

In many of the cases where NARS disseminate the results of NGO research, however, these research results are not technological but methodological. As we noted earlier, the development of methodologies is a particular strength of many NGOs. Thus, for instance, some of the FSR approaches developed by RKM, and disseminated in and beyond the E. India Rainfed Farming Systems Research Network, have stimulated the adoption of FSR by parts of the NARS.

In Thailand, the methods developed by ATA for farmer-to-farmer dissemination of fish farming technologies have been taken up by extension services (Jonjuabsong and Hwai-Kham 1993). In Kenya the Forest Department took up extension methods developed by NGOs (Mung'ala 1993), and in Ghana, interaction between NGOs (especially ACDEP and TAAP) with Nyankpala Agriculture Research Station has stimulated the latter's experimentation with farming systems and participatory methods (Amanor and Wellard 1993).

Box 5.4 Research on soya by the Mennonite Central Committee in Bangladesh

The MCC is a US-based volunteer organization which has been working in Bangladesh since 1970, and has produced comprehensive annual reports on its agricultural research work since 1975. Soya is not a traditional indigenous crop in Bangladesh. Twenty years ago, however, it appeared to offer the potential to fit into farmers' cultivation cycles as a low-cost highly nutritious crop. The Bangladesh government set up the Co-ordinated Soyabean Research Project involving seven GOs, a food corporation and MCC. However, the project failed to identify a variety capable of being stored as seed in humid conditions from one year to the next. MCC persevered with varietal screening trials, in 1985 settling on an Indian variety which met local requirements. MCC also worked with private snackfood companies on market development and processing technologies. Healthy demand had led to the cultivation of over 500 ha by 1989. Despite some setbacks – the National Seed Board could, for instance, not be persuaded initially to agree the official 'approved release' of this variety – the crop was included in the government's Crop Diversification Programme in 1989, and soya became part of the Ministry's overall crop promotion strategy.

Source: Buckland and Graham 1990

NGOs disseminate the results of their own research

The final column of Table 5.1 indicates situations in which NGOs develop and disseminate technology in isolation from the NARS. They do so for two main reasons:

- The resource constraints faced by NARS may simply mean that they cannot afford to cover either the commodities or the agroecological areas in which the NGO wishes to work;
- NGOs' visions of the future of the rural poor, and the strategies they pursue in support of these visions, may be different from those of government (many of these are the 'agroecological' NGOs discussed in Chapter 3).

NGO responses to the former case include: El Ceibo's research on cocoa in Bolivia (Box 3.7; Trujillo 1991; see also Bebbington and Thiele 1993); CRS's work on production and processing technologies for sesame in the Gambia (Owens 1993); the work of several NGOs on animal traction in the Gambia (Sarch and Copestake 1993), and GVAM's (Gwembe Valley Agriculture Mission) work on vegetables in Zambia (Copestake 1993). As examples of NGOs with visions that differ from those of the NARS, ENDA and other

NGOs have been collecting millet and sorghum genetic material in Zimbabwe in an effort to 'restore' these crops to their previous importance in areas where recently-introduced maize is threatened by drought (Chaguma and Gumbo 1993). CARE in Bangladesh (Dean 1990), IIRR in the Philippines (Gonsalves and Miclat-Teves 1991) and CIED in Peru (Guerrero 1991) have all found that their organic or agroecological perspectives require research which government has not yet addressed. FIVDB (ducks) (Nahas 1993) and BRAC (tubewells owned and managed by the landless) (Mustafa *et al.* 1993) have both found that the Bangladeshi NARS attaches low priority to technology development for the landless and for women, who constitute major components of NGOs' clientele.

The only case that does not fit entirely in to these two categories is that of BAIF – a research foundation in India (Satish and Farrington 1993). While BAIF's vaccine production programme responds to a gap in the market – the private commercial sector sells imported vaccines at prices unaffordable by small farmers, and the response of government research has been inadequate – its programme of artificial insemination for cross-bred dairy cattle began as a pioneering effort to fill an area under-researched by the NARS, but has developed into a large-scale service delivery programme which is capable of full cost recovery and of competing effectively with the private commercial sector.

LINKAGE MECHANISMS SUPPORTING INSTITUTIONAL ROLES

If different organizations are performing tasks in different sub-systems of the ATS, it is likely that there will be a range of actual and potential interdependencies among them. Merrill-Sands and Kaimowitz (1991) have identified some of these interdependencies in the case of public sector research and extension agencies within a NARS (Box 5.5). Many of these apply also to interdependencies between NARS and NGOs, but an important difference in institutional mandates should be noted: different agencies within a NARS – for example, the research and extension branches – are *required* to work with each other by virtue of being under the same public sector umbrella; conversely NGOs and NARS will only work with each other if there is a clearly perceived advantage in doing so. Furthermore, an NGO has much greater flexibility than does, say, the adaptive research branch of a NARS in forging links with other agencies (such as an international NGO network) if it feels unable to meet its needs through contact with in-country agencies.

For these reasons, inter-organizational linkages between NGOs and NARS, and particularly if they are to be the fully collaborative type, do not simply look after themselves.

Conscious efforts have to be made to design joint activities, to allocate responsibilities and to provide resources in support of them. These different

136

aspects of inter-institutional co-ordination we term linkage *mechanisms* and we turn to analysis of how they have been managed in the next section (cf. Merrill-Sands and Kaimowitz 1991).

Box 5.5 Interdependencies between actors carrying out research and technology transfer

---

Researchers need the support of technology transfer agents in:

- Selecting the right technology for development;
- Appropriate design of trials;
- Identifying representative locations for trials;
- Providing feedback about farmers' reactions to new technologies;
- Interpretation of trial results.

Technology transfer workers need the support of researchers in:

- Providing information about the characteristics of new technologies;
- Explaining the context in which new technologies should be used;
- Ensuring that the messages being delivered to farmers are correct and sufficiently detailed;
- Providing specialist services such as soils testing and pest identification.

Source: Merrill-Sands and Kaimowitz 1991: 16

Types of linkage mechanism

Our use of the terminology of linkage mechanisms is based on ISNAR's work (Merrill-Sands and Kaimowitz 1991). While ISNAR's analysis dealt only with public sector organizations, it is amenable to the case material in this study. Furthermore, using it here allows a degree of complementarity between the two studies.

As we will see, there are a range of mechanisms, and the choice of the most appropriate mechanism depends on the functions performed by each actor as well as on a range of contextual factors. Both vary considerably, so it is no surprise to find a wide range of mechanisms in use (Table 5.2).[10] These can be either (1) *operational*, supporting project-type implementation of specific aspects of ATD, or (2) *structural*, where linkages are at an institutional level between NARS and NGOs and give each institution some influence over the resource allocation and strategic decisions of the other. Each of these is considered in turn as we now discuss Table 5.2 in detail.

Experiences with linkage mechanisms: an overview

Table 5.2 shows the different linkage mechanisms which were found in the case studies, distinguishing those described by NGOs and those by NARS. The

Table 5.2 Structural and operational linkage mechanisms found in the case studies

	Import-ance	Formality (mainly formal or informal)	Structural linkage mechanism	Operational linkage mechanisms — Joint professional activities	Whether specific resources allocated in support of linkage
AFRICA					
(a) NGO cases					
Machakos NGOs	*	Inf		NGOs disseminate soil and water conservation technologies	No
Silveira House	*	Inf		NGO disseminated hybrid maize	No
GVAM	*	Inf		Adaptive trials on GO grain varieties	Yes/No'
ACDEP/ Langbensi/ TAAP	*	Inf, F	NGO members on advisory board of NARS	Train GO staff in participatory methods; Adaptive varietal trials; GO scales up animal traction	No
Rodale/CRAR	**	F		Joint trials on organic matter utilization	Yes
ARAF	*	F		Joint trials on composting and organic matter utilization	Yes
(b) Public sector cases					
KEFRI	***	F		Joint agroforestry trials	Yes
FITT	**	F		NARS scales up NGO extension method; Joint crop trials co-ordinated with numerous NGOs	Yes
ASIA					
(a) NGO cases					
RDRS	**	F		Government issued tenders for manufacture of NGO technology	No
BRAC	**	F		NGO organized communities to take advantage of government supplied inputs; Joint NGO–NARS training	Yes
Proshika	**	F		NGO organized communities to take advantage of government supplied inputs	Yes

Case		Type	Institutional linkage	Activity	Yes/No
MCC	*	F		Joint trials on fruitfly trapping methods	Yes
PRADAN	*	Inf		GOs disseminate NGO soya varieties	No
AWS	***	F		NGO 'demystifies', scales down and disseminates NARS technologies to groups it has organized; Joint design and implementation of IPM strategies; Joint design of IPM bulletin	Yes
RKM	**	F	NARS staff on Management Committee of FSR Project	Trials with NARS and university on chilli varieties	Yes
AKRSP	**	F		Joint design and contributions to FSR Newsletter	Yes
BAIF	*	F		Design of participatory training methods with and for NARS trainers; Forage research and animal disease monitoring as part of all-India research network	Yes
MBRLC	***	F	NARS scientists on NGO Board	Agricultural development projects take up NGO land management technology; Training of GO staff	Yes
ATA	**	Inf		NGO devises farmer-to-farmer methods for disseminating GO technology; NARS subsequently adopts dissemination method	No
(b) Public sector cases					
ERDB	**	F		Adaptation of integrated livestock model	Yes
Surin Fisheries	*	F		Trials with rice-fish culture	Yes
LATIN AMERICA **(a) NGO cases**					
CESA-Ec	**	F,Inf	Committee in rural development programme	Technical advice from NARS for adaptive trials phase I and II; Co-ordination for adaptive trials phase II; Co-ordination for transfer in fruit	Yes
CAAP	*	Inf		General agreement for technical co-operation	No
FUNDAEC	**	F, Inf		Assisted NARS in locating trials; Collaborative livestock project for investigation and transfer	No
SEPAS	**	F, Inf		Diffusion of technology	No
IDEAS	*	Inf		Visits by NARS researchers	No
CIED Cajamarca	*	Inf		Study visits to NARS	
CESA-Bol	***	F		Agro-ecology symposium with NARS	Yes

Table 5.2 Structural and operational linkage mechanisms found in the case studies (continued)

	Import-ance	Formality (mainly formal or informal)	Structural linkage mechanism	Operational linkage mechanisms	
				Joint professional activities	Whether specific resources allocated in support of linkage
CIPCA	***	F	Participation in annual planning of NARS jointly run experimental station	NARS field days NARS bulletins NARS courses	Yes
CEIBO	**	Inf	Inter-institutional co-ordination Bimonthly meetings NGOs and NARS	Consultations about coffee with NARS NARS trained extensionists	No
PROCADE/IBTA	*	F	Mutual representation on governing bodies	Co-ordination for forage research	Yes
GIA	***	F,Inf	Ex GIA staff join NARS Joint committee	NARS contracted NGO for extension Carried out diagnosis for NARS NGO trained NARS	Yes
AGRARIA	***	F,Inf		Joint preparation of research project Joint trial NARS contracted NGO for extension	Yes
(b) Public sector cases					
CIAT	***	F	Zonal meetings with NGOs; Joint consultation on operation of Technology Transfer Unit.	Joint diagnosis NGOs participate in adjustment trials Joint workshops to determine recommendations Extension bulletins for NGOs Information centre for NGOs Field days for NGOs	Yes
INIAP-PIP	**	Inf		Joint diagnosis Exchange of technical information NARS supplies seed to NGO Joint adjustment and transfer Joint planning of demonstration plots NGO advises about adjustment trial location	No

* ittle importance; ** some importance; *** important.

F formal mechanisms predominate; Inf informal mechanisms predominate.

Notes: 1 Initially, GVAM provided a motorcycle to facilitate visits to project areas by the NARS extensionist. Arrangements were subsequently made for him to live in housing provided by the project. However, he found this too remote, and so resource-sharing arrangements ceased.

first column of the table indicates the importance attached by the case study institutions to the linkage mechanism concerned. This is high in Bolivia and Chile, where, as the third column indicates, structural linkage arrangements have been introduced as a consequence of public sector reforms that have influenced the NARS (see above). Elsewhere, links have been particularly important to the Kenya Forestry Research Institute insofar as it gained considerable strength in its early years by drawing on NGO field experience (Mung'ala 1993). Links have also been an important part of the strategy of MBRLC, which feels that government projects which scale up its technology do make valuable contributions to making hill-slope agriculture more productive and more sustainable (Watson and Laquihon 1993). In the case of AWS, it is clear that its programme of IPM (Integrated Pest Management) work could not have taken off without technical and material support arranged through the NARS (Satish and Vardhan 1993).

The second column of Table 5.2 indicates whether formal or informal linkages are relatively more important. Formal linkages were important in only four cases in Latin America, and all were in Bolivia or Chile – once again reflecting the recent public sector reforms there. The relative lack of formalized linkages might be attributable to three factors: a predisposition towards informal links arising from the relatively small size of research communities in most countries; the similar social backgrounds of NGO and NARS staff, many of whom may share friends or university experiences; and the difficulty of getting formal agreement approved through what are still often inefficient and centralized bureaucracies. In Africa and Asia, by contrast, formal linkages were predominant, and in most cases specified the resources to be contributed from each side.

Formalization can have its drawbacks, and as NGOs' experience with INDAP in Chile illustrates, the level of detail in some formal contracts can be excessively rigid and often very imperfect (Box 5.6). However, informal agreements leave each side in a particularly vulnerable position when it comes to requesting the fulfilment of verbal promises (Box 5.7 – INIAP-PIP), and tend to be nullified when key staff change.

Structural linkage mechanisms

Three different types of structural linkage mechanism were found in the case studies:

1 Co-ordination units, which have the specific function of co-ordinating activities between the two actors;
2 Permanent committees made up of representatives of the NARS and the NGOs;
3 Representation of one organization in the executive body of the other.

Structural linkage mechanisms are necessarily formal, and institutionally

---

Box 5.6 Chile: INDAP contracts NGOs to carry out extension

In its Programme for Technology Transfer, the part of the Chilean NARS responsible for technology transfer (INDAP) provides extension to farmers through a number of 'private consulting companies'. The activities of the companies are financed by a government subsidy of about $US330 for each subsistence farmer who participates. INDAP plans the allocation of resources and monitors the extension programme.

Following the restoration of electoral democracy and civilian rule in 1990, INDAP began to allow NGOs to participate as consulting companies. While NGOs have criticised various aspects of the Programme on the grounds that it is:
● Designed for individual farmers;
● Covers only technical assistance and is not related to credit programmes;
● Top down;
● Short term.

Many of them have none the less chosen to participate in it: in part to influence it, and in part to gain access to funds. AGRARIA presented itself as a consulting company and, by 1991 has seven separate INDAP programmes (this has since increased). Extension methods were based on prior experience of AGRARIA in the region. Although the Programme has become more flexible, its rigidity still affects the way extension is carried out. None the less, since the state is providing resources for transfer activities AGRARIA feels it can no longer countenance funding transfer work with its own funds.

GIA also decided to enter the Programme, and several of its technical teams have formed themselves into a consulting company. Its motivations for doing so are, however, slightly different. GIA's objective is to study the Programme's operation and analyse its impact on peasant farmers. On the basis of these results, it will suggest alterations to the Programme's design. Results and recommendations will be passed on to INDAP.

GIA took charge of three multi-modules in the Programme. Involvement in implementation has helped it both to experiment with ways of achieving a gradual transformation in the Programme and to pass on its findings to INDAP through regular contacts. For example, GIA (like AGRARIA) has made an attempt to carry out extension with groups and not individuals: on the basis of that experience, changes have been recommended to INDAP, and thence to other consulting companies by INDAP.

Sources: Aguirre and Namdar-Irani (1992) and Sotomayor (1991)

---

recognized. They are also usually permanent because they seek sustainable co-ordination – though they vary in their ability to impose decisions.

The third column of Table 5.2 indicates the cases where structural linkage mechanisms are in place. Few examples are found in Africa and Asia: they include NARS representation on the board of BAIF in India (Satish and Farrington 1993), and on the FSR programme management committee of RKM (Chakraborty *et al.* 1993) and, in Ghana, NGO representation on the advisory group of Nyankpala research station (Alebikiya, pers. comm.). An Indian NGO, PRADAN, has representatives of the public sector, academic and the private commercial sector on its general governing body, whose mandate is

---

Box 5.7 INIAP-PIP: advantages and limitations of informal linkage mechanisms

In the late 1970s the National Institute of Agricultural Research (INIAP) of Ecuador created a Programme of Production Research (PIP) as an on-farm, systems-based research programme. Each PIP Unit carries out on-farm trials for selection and adaptation of technology. The farmer participates in the preparation, setting-up and harvest of the trials. The intention is to transfer trial results to other parts of the NARS and to extensionists, and to generate feedback on farmers' needs in order to orient the work of INIAP's experimental stations.

Two of the ten PIP Units (Chimborazo and Cayambe) had worked informally with NGOs. In Cayambe, INIAP worked on separate occasions with CESA and CAAP, building on existing friendships between members of the different institutions that originated from university days. They discussed technology options with each other, and PIP staff provided the NGOs with improved seeds which they then jointly tested and disseminated. A strong work ethic on both sides, and flexible working practices, made possible a high degree of collaboration.

Active small farmer participation in the work of the PIP Units was made possible both by the work of the NGO staff, who organized farmers for field activities, and the efforts of the PIP staff to reach the remote *campesino* communities where the NGOs worked.

However, the most fundamental limitation on these collaborations was the lack of any formal agreements, which made it impossible to demand the fulfilment of any commitments that were made. This meant that when, after about three years, the staff of the NGOs changed, collaboration between the PIP Units and the NGOs ceased.

Source: Cardoso 1991; interviews

---

to formulate policy guidelines (Vasimalai 1993). By contrast, in Latin America there is more evidence of structural linkage mechanisms, some of which have been introduced as a result of recent radical change in management and orientation of the ATS following public sector reform and the election of democratic governments (see above). Two types of structural linkage mechanism – permanent committees and co-ordination units are examined below in more detail.

The permanent committee

The permanent committee is made up of representatives from various institutions. The clearest example of a permanent committee which functions as a linkage mechanism is in Chile where, in 1990 INIA (the NARS) began to establish over fifty Centres for Adjustment and Transfer of Technology (CATTs), each corresponding to an agroecological zone. The Committees that monitor and advise on the CATT's activities include representatives of NGOs, NARS and consulting companies.

Furthermore, the design of the CATT programme (another post-Pinochet change in Chile) owes much to the influence of NGOs. NGO staff were involved in the initial consultancies, and subsequently NGOs have also had institutional influence over the programme. For instance, through formal links with the NARS, GIA contributed to the design of the CATT Programme at three levels: at a *national-level*, it contributed to the design of the main methodological elements; at a *regional-level*, it worked in the identification of broad agro-ecological areas in which CATTs should be located in the VIII Region; and at a *local-level*, it concentrated on the practical aspects of establishing three CATTs, which included the selection of plots, the identification of research topics, and the design of specific methods of operation (Sotomayor 1991; see also Bebbington and Thiele 1993).

The co-ordination unit

Co-ordination units may form part of one or other of the institutions involved in the link or may be independent bodies (Merrill-Sands and Kaimowitz 1991: 41). Considerable experience has already accumulated on co-ordination units (or positions) which have been set up to enhance research-extension linkages *within* the public sector, for example in Nigeria (Ekpere and Idowu 1989) and Zambia (Kean and Singogo 1990).

The clearest example of a co-ordination unit found in the case studies is the Department of Technology Transfer (DTT) of CIAT in the Santa Cruz region of Bolivia (Boxes 5.8, 5.9). This was formed with the express objective of establishing linkages between CIAT's research activities and the extension work being carried out by NGOs, producers' organizations, commercial companies and public institutes in the area CIAT was mandated to serve.

The localized nature of NGO activities has in some cases led them to group together in networks, or form co-ordinating bodies. When these explicitly seek to develop relations with the public sector they can also be classified as co-ordination units, even though they may also have other objectives (such as policy negotiations). For instance, PROCADE is a programme of inter-institutional co-ordination for the ATD activities of twelve Bolivian NGOs. It has acquired some of the functions of a co-ordination unit, both within the NGO sector, and between NGOs and IBTA, the NARS. Among its members, PROCADE co-ordinates the planning (and funding) of research (aiming to avoid duplication), and the inter-institutional exchange of information. At the same time, PROCADE is the NGO representative on the co-ordinating council of IBTA, where it assumes a co-ordinating role, as well as the role of policy critic. However, PROCADE's success in doing this is itself hampered by jealousies among its members, and by their own poor record in inter-institutional co-ordination at the level of each individual NGO. Aside from illustrating the difficulty of promoting co-ordination among NGOs, this experience also shows that establishing a structural linkage mechanism is not in itself sufficient to establish functioning links. It is necessary to complement it with specific operative linkage mechanisms.

144

The limits to inter-institutional co-ordination through structural linkages

In Chile and the Santa Cruz region of Bolivia structural linkage mechanisms have been introduced and have allowed better functioning of operational linkage mechanisms. However, the success achieved in these cases depends fundamentally on both the wider political and institutional conditions, and the local factors which have favoured NGO–NARS linkage (Velez and Thiele 1991; Bojanic 1991; Sotomayor 1991). In Santa Cruz, the fact that the public sector was strong, and there had not been severe repression in the region in earlier years facilitated good NGO–NARS relationships. Perhaps most important though was the network of very strong inter-personal contacts that cut across institutional boundaries, *and* the fact that there was some convergence between NGOs and the NARS on agricultural development strategies (Box 5.9). In Chile, the re-democratization of 1990 created a climate in which many NGO staff moved into the NARS, similarly creating a situation in which there was greater convergence of opinion over development strategies, and a network of personal contacts cutting across institutional boundaries.

In other places it has been more difficult to introduce robust linkage mechanisms. For instance, in the case of CESA and INIAP in Ecuador (CESA 1991), as differences of interest between the NARS and the NGO gradually became clear, CESA reached the conclusion that only very specific linkage mechanisms should be implemented in situations where there was a clear coincidence of interest. The limited success of structural linkages in Ecuador has been attributable to imprecise objectives, inadequate designation of responsibilities and budgetary allocations, inadequate mechanisms for monitoring and evaluation, the rapid staff turnover in INIAP and its predominant export crop mandate.

Operational linkage mechanisms

Adapting a classification developed by Kaimowitz and Merrill-Sands (1991), we identify two broad types of operational linkage mechanisms:

1 Joint professional activities;[11]
2 Resource-allocation procedures.

The principal mechanisms identified in each category are shown in Box 5.10. They can be formal or informal, mandated or voluntary, permanent or temporary (Kaimowitz *et al*. 1990: 233).

Joint professional activities

Joint professional activities in the third column of Table 5.2 embrace joint diagnosis, planning, programming and review, joint evaluation, the joint

145

Box 5.8 Linkage mechanisms in CIAT (Bolivia)

CIAT Department of Technology Transfer: complementing a co-ordination unit with operational linkage mechanisms

Public sector extension services in eastern Bolivia have long been characterized by chronic weakness (Thiele *et al.* 1988). Under a new strategy devised in 1989, CIAT established a co-ordination unit – the Technology Transfer Department (DTT) – whose role was not to work directly with farmers, but with various intermediate users (IUs) of technologies who had their own local extension services. NGOs are one of the most important types of IU.

The DTT has subject-matter specialists and zonal specialists whose work is supported by a communications section. The subject-matter specialists (SMSs) are in regular contact with their corresponding CIAT researcher and collaborate on some research work. They package research information for delivery to IUs and are mandated to transmit feedback on farmer needs to the researcher.

Frequently, technologies developed in the experimental centres are still not ready for transfer. SMSs therefore carry out on-farm adaptive trials, in addition to ensuring that extensionists pass on the appropriate messages to farmers. Other duties of the SMSs include the preparation of technical bulletins for extensionists, the enhancement of feedback and advice to extensionists (for instance, on the establishment of demonstration plots) and on how to give talks to farmers.

CIPCA and CIAT jointly manage a local experimental station

When CIPCA began to work in the southern hills of Santa Cruz Department in 1976, no research institution was catering for that area, and CIPCA decided to establish its own experimental centre for research in fruit, pigs, cattle, and grain crops. Subsequently, CIPCA reached agreement with CIAT for the joint management of research at the centre, and in some respects the centre served as a model for six further regional research centres that CIAT established in order to adapt technology to local agroecological conditions elsewhere in Santa Cruz. CIAT provided a full-time researcher in the regional research centre and operating costs were shared with CIPCA. Extension is carried out solely by CIPCA.

The successful resolution of early difficulties arising from lack of co-ordination between CIAT and CIPCA in management of the researcher led to agreement that the centre as a whole should be jointly managed. A formal agreement to this effect has been signed, specifying inputs from both sides, which has formed the basis for numerous informal exchanges and has led to a sharper focus in CIAT on issues relevant to the area.

Source: Bebbington *et al.* 1992: 53; Garcia *et al.* 1991

CIAT and CESA-Bolivia: Non-permanent mechanisms for co-ordinating adaptive research

In the absence of institutions in CESA's area of operation which could provide advice on agroecological methods CESA began to establish contacts with different institutions involved in tropical agriculture in other parts of Bolivia. Informal links with CIAT and its British Tropical Agriculture Mission support team led to visits by experts for brief periods to advise the field team and to give short training courses. Relations with CIAT were also very important in the supply of seeds and tree seedlings for use in agro-silvipastoral systems.

In addition, groups of *campesino* promoters from the project made study visits to the research stations and experimental plots of CIAT and IBTA. This created an informal network of contacts and enormously enriched the body of agroecological knowledge underlying the project. In turn, CIAT and IBTA benefited from being able to observe the performance of new agro-ecological activities in CESA's field location.

In order to improve the co-ordination of these activities and the exchange of information generated, CESA and neighbouring institutions created a local network which organized an agroecological symposium attended by the external experts and other local institutions. The symposium served to consolidate local thinking about agricultural development strategies, and provided a base for subsequent project design in CESA.

Source: Kopp and Domingo 1991

Box 5.9 Factors favouring CIAT's relations with NGOs in Santa Cruz

Social factors

- History of less political violence in Santa Cruz;
- Strong personal relationships between CIAT and NGO staff;
- Movement of staff between NGOs and CIAT.

Institutional factors

- CIAT was strong and donor funded in key areas (e.g. agroforestry research);
- CIAT was partly protected from political interference.

Technological factors

- CIAT and NGOs share opinions on the 'appropriate technology' for poor farmers;
- NGOs recognize need for technology research;
- Strategic gaps in farmer technical knowledge.

Source: Bebbington and Thiele 1993

---

implementation of applied and adaptive research, and of dissemination. Each is considered briefly in turn:

Joint problem diagnosis

Several instances of joint NGO–NARS problem diagnosis are found in the case studies: in Chile, GIA carried out diagnosis on behalf of the NARS (Sotomayor, 1991); in Bolivia, NGOs joined CIAT in the design and

---

Box 5.10 Operational linkage mechanisms

Joint professional activities

Joint problem diagnosis
Joint priority-setting and planning exercises
Formal collaboration in trials, surveys, and dissemination activities
Joint decision making on release of recommendations
Regular joint field visits
Informal consultations
Publications, audio-visual materials and reports
Joint training activities or seminars

Resource allocation procedures

Formal guidelines for allocating time for collaborative activities
Specific allocation of funds for collaborative activities
Staff rotation and secondment

Source: Adapted from: Merrill-Sands and Kaimowitz 1991: 52

---

implementation of *sondeo* (rapid survey – literally 'sounding') methods (Velez and Thiele 1991); in India, AWS and NARS institutions were jointly involved in identifying pest problems on castor and in designing past management strategies (Satish and Vardhan 1993); in Gujarat, AKRSP identified jointly with NARS staff, a number of fundamental problems in the latter's approach to farmer training and jointly designed ways of enhancing its relevance (Shah and Mane 1993). In Kenya, CARE, Mazingira, KEFRI (Kenya Forestry Research Institute) and ICRAF (International Centre for Research in Agroforestry) in the early 1980s, conducted a number of joint diagnoses which influenced the design of subsequent research (Buck 1993; Mung'ala 1993). In Senegal, Rodale/CRAR and ARAF, both working together with the NARS (ISRA) jointly identified low soil organic matter content, high fertilizer costs and irregular supplies as limiting factors in agricultural production, and designed experiments on the basis of this (Diop 1993; Dugue 1993).

Joint planning

Joint planning requires formal mechanisms so that decisions taken have the full support of the institutions involved. One of the few Latin American examples from the case studies of formal planning mechanisms – none the less a highly innovative one – is the participation of NGOs in research planning meetings of CIAT in Bolivia (Velez and Thiele 1991).

By contrast, joint planning agreements have been formalized in the majority of cases reported from Africa and Asia. Most of this joint planning relates to the design and location of research trials. Variants on the theme include:

148

BRAC's and Proshika's projects jointly planned with line departments which involve the NGOs in the design of community organizational models appropriate for the implementation of improved animal production (discussed in Box 3.9), the joint design of participatory training methods for use by government training staff (AKRSP – Box 5.12), and the joint specification of tenders to be offered by government for the manufacture of NGO-designed treadle pumps (Box 5.14).

Joint programming

Programming deals with the organization of specific activities over time. In two of the Latin America case studies (El Ceibo and CIAT) programming was carried out in meetings in the area where NARS and NGO field staff worked. This kind of meeting served to build on complementarities between activities which had already been planned. It should be considered as a supplement to longer term planning and not an alternative.

In a number of other cases, project-based linkages were organized to allow periodic reflection and planning for the forthcoming period. For instance, the AWS experience with integrated pest management involved particularly complex arrangements for the provision of inputs which had to be reformulated periodically.

Joint evaluation

In the absence of joint evaluation, feedback on the performance of operational linkages is likely to remain weak. No examples were found in the African or Latin America case studies where NGOs and NARS had jointly carried out evaluation. However, in Asia, the Proshika (livestock) and BRAC (poultry) cases discussed elsewhere in this volume (p.80; Box 3.9) have been evaluated. A particular concern arising from these evaluations is the extent to which government services responsible for providing vaccines will remain accessible to the local membership groups that have been formed in each case. Evidence is already emerging of government wishes to economize on these services, and so lengthen required travelling distances for group members. Also in Asia, research in which BAIF is involved as part of two All-India Co-ordinated Research Programmes (in forage research and animal disease monitoring) is monitored and evaluated by joint teams of scientists in accordance with India Government norms (Satish and Farrington 1993).

Joint implementation of applied and adaptive research

As Table 5.2 indicates, a large number of cases of joint research were found among the case studies, only a few of which can be highlighted here.

Links across applied and adaptive research were found in the ARAF and

Rodale/CRAR case studies from Senegal, in which the results of applied research by ISRA (the NARS) on soil and water conservation were used in field trials designed to identify locally appropriate composting practices (Diop 1993; Dugue 1993). In Bangladesh, applied work by the NARS to select appropriate types of fruitfly bait was put to use with MCC in field trials to test alternative designs of trap (Buckland and Graham 1993). In W. Bengal, irradiation techniques carried out in a NARS institute produced a wide range of genetic material in chilli, which was then field tested in a joint project between the NARS, a university and RKM (Chakraborty et al. 1993).

Numerous examples are available of professional links in adapting technology to the socio-economic characteristics of specific clients. For instance, CESA-Bolivia developed special linkage mechanisms to support its adjustment trials. The mechanism was formal but not permanent, consisting of field visits by the NARS staff and study visits by CESA's 'promoters' to the NARS (see Box 5.8). This mechanism supported adjustment by the NGO of technology developed by the NARS.

In The Gambia, under the FITT programme, numerous NGOs field-tested varieties of staple grains which had been selected by the NARS. Advice was given by the NARS in the design of trials, but resource limitations prevented adequate levels of support and the responsibilities of each side were not spelled out and scheduled clearly enough (Box 5.11). In other work in The Gambia, NGOs arranged for farmers – both individuals and in groups – to produce good quality seed by multiplying up suitably isolated plots of new seed provided by the NARS, which then provided seed testing and certification facilities (Henderson and Singh 1990).

Joint dissemination and training activities

Three cases were found in S. America (CESA-Ecuador (CESA 1991), IDEAS (Chavez 1991), INIAP-PIP (Cardoso 1991)) where there had been some loosely specified collaboration in dissemination. Only in Bolivia and Chile did well-developed mechanisms exist.

In Bolivia the mechanism is voluntary in the sense that the NARS Department of Technology Transfer (DTT) (Box 5.8) invites extensionists of different organizations, including NGOs, and researchers to give cropping recommendations for publication in technical bulletins. It also publishes technical material which takes into account the needs of NGO extensionists, runs training courses and field days with and for them, and has an information centre which they can access.

Chile has both mandatory and voluntary mechanisms. As an example of the former, INDAP contracts NGOs to provide technical assistance to farmers. As Box 5.6 indicates, the NGOs have realized that this mechanism requires modification and are trying to improve it. In the latter, GIA, for instance, has trained NARS staff in methodologies for problem diagnosis, production

---

5.11 NARS-initiated linkages in adaptive research: the Farmers' Innovation and Technology Testing Programme in The Gambia

---

The FITT programme was launched in 1989 under the Gambia Agricultural Research and Diversification Project (supported by the United States Agency for International Development) in response to calls for an increased flow of technologies from the NARS to farmers. It aimed to involve the substantial resources of NGOs in The Gambia, together with public sector extension, as intermediaries between research and farmers in the on-farm testing and adaptation of technologies. The launch was marked by an orientation and planning workshop organized by the NARS, in which seven NGOs participated. A total of eight NGOs participated in on-farm trials under FITT in the first season, but by the third this had declined to only two. The absence (abroad) of the director of agricultural extension at the initial planning stages contributed to a weak commitment by this part of the NARS throughout.

The rapid decline of a functional relationship which has strong intuitive appeal (it shares, for instance, the same philosophy as that on which the CIAT (Bolivia) technology transfer department was established – Box 5.8), and which was initially taken up with enthusiasm, has been attributed to shortcomings on both sides. A 1992 survey of four of the original NGO participants and their clients (Sarch 1993) suggests the following problems on the NGO side:

• Inadequate participation with farmers in selecting the technologies to be tested and in agreeing management practices;
• Unrealistically high levels of inputs provided by the high proportion of NGOs who were ideologically predisposed to communally-managed trial plots;
• Where trials were conducted with individual farmers, these were predominantly male and of higher than average income.

The following shortcomings were identified on the NARS side:

• Late delivery of essential inputs for trials; poor quality of seeds. These factors resulted in such late planting of nine trials in five villages that little yield was obtained;
• Inadequate allocation of responsibility between NARS and NGOs for collection of materials for trials;
• Rapid attrition of NARS staff during the programme period.

Overall, three of the eight villages surveyed rejected at least one technology completely, and one rejected all four that it tested. At the other end of the spectrum, one technology has been widely adopted in the trial village. In two further villages, farmers are experimenting further with seed saved from the limited yield obtained from late planted crops.

In institutional terms, the two NGOs remaining in the programme in the third year seem to value links with the NARS and appear set to continue. With more careful design of the linkage, this proportion would undoubtedly have been higher.

Source: Sarch 1993

---

systems research, and the analysis of peasant rationality and peasant decision making and organization (Sotomayor 1991).

In India, AWS and a NARS institute jointly designed extension material on integrated pest management, and staff from RKM, from NARS institutes and from universities all contribute to the newsletter produced by RKM for the E. India Farming Systems Network. RKM trains some 6,000 extensionists annually, many of them government staff, in a wide range of issues including systems approaches (Chakraborty *et al.* 1993).

A particularly illuminating joint effort in training and dissemination is provided by AKRSP's work in Gujarat, India (Box 5.12). In order to make hitherto 'top-down' farmer training courses given by government institutions more participatory, AKRSP first helped farmers to develop a capacity for identifying the opportunities and constraints that might be met by technological change, and for conducting and sharing the results of small-scale experimentation. Once this process had become established, NARS staff were brought in to learn from it and to identify how their approaches to training might be modified and so replicate it elsewhere.

Box 5.12 AKRSP re-trains government trainers in India

AKRSP's philosophy has been to draw on the large number of specialist government training institutions to provide technical courses for farmer-clients of its own agriculture programmes. Farmers' evaluations of early efforts by GO trainers were critical of the 'blueprint' courses offered: they were insufficiently adapted to local agroecological conditions, failed to draw on relevant experience gained by farmers, spent too much time in the classroom and not enough in the field, and did not allow for 'try it and see' approaches.

AKRSP's response came in two stages. The first was to strengthen farmers' capacity for identifying their own opportunities and constraints through resource inventories. These were led by extension volunteers trained by AKRSP, and led to the development of village natural resource management plans, and to the identification of on-farm experiments that were conducted to meet specific needs.

In a second stage, once these processes of analysis, prioritization and experimentation had become consolidated, AKRSP began to 'hand over' by familiarizing GO trainers with the participatory approaches developed, and building up their own capacity to introduce them elsewhere. Realizing that the radical reform of hitherto top-down training methods implied by this approach might cause resistance within GOs, AKRSP organized 'exposure workshops', in which senior administrators from all thirty-four training institutes in Gujarat State, together with a number of senior state-level officials, were given demonstrations of the participatory approaches that had been developed, and heard farmers' criticisms of earlier courses.

As part of the decision formally to take up participatory approaches to training, it was agreed as a result of the workshops that each GO training institute would initially 'adopt' one to three villages to experiment with ways of enhancing productivity in a practical, participatory fashion, and that small thematic networks should be established across the institutes as a means of professional support to staff.

Source: Shah and Mane, 1993

In Kenya, good working relations between the Forestry Extension Services Division and a number of NGOs have led to joint programmes of extension and training, and to efforts to harmonize a range of activities (Box 5.13).

Box 5.13 The Kenya Forestry Extension Services Division (FESD) and links with NGOs in training and extension

The FESD, prior to 1989 known as the Rural Afforestation and Extension Scheme, has developed strong links with a number of NGOs, and has sought to draw on their experience to stimulate participatory approaches to extension and service provision. Its change in policy from large, centralized to small farm-based seedling nurseries has, for instance, largely been stimulated by NGOs' experience. Together with CARE in Siaya District it has jointly participated in agricultural shows and organized competitions for farmers and school children. The two organizations were among several NGOs and GOs that in 1990 organized the first National Agroforestry Extension Training Workshop. Together with KENGO (Kenya Energy and Environment Organizations), the FESD has participated in joint training seminars on seed collecting, handling, storage and distribution, land-use systems and indigenous trees. The FESD also works with NGOs to draw up guidelines for the pricing of seedlings according to varying conditions across the country, in an effort to prevent any organization from introducing excessively subsidized prices and so undermining the operations of others.

Source: Mung'ala 1993

Resource allocation procedures

The allocation of resources, by one or other partners, is a prerequisite to the formation and development of linkages. In general, informal linkages imply the *ad hoc* provision of staff time and occasionally such other resources as transport and facilities for meetings. By contrast, formal linkages require pre-planned resource allocations. Careful planning of responsibilities is particularly important in fully collaborative efforts – such as joint on-farm trails – in which a successful outcome relies on carefully scheduled inputs from each side. The experience of FITT in the Gambia (Box 5.11) illustrates the types of problem that can occur when responsibilities are either unclearly defined or are not adhered to.

Certain types of sequential linkage also involve a degree of formality. Substantial resource allocations were made by the Bangladesh government, for instance, in order to scale up the treadle pump technology developed by RDRS. While RDRS's research and dissemination activities formed part of the same technology path, and so are clearly linked in this sense, resources were allocated by government for manufacturing only, and not in support of any direct interaction with the NGO at the scaling up stage, which led to much subsequent waste of resources (Box 5.14). Overall, the design of formal resource allocation procedures requires time and effort, and while informal procedures have the attraction of flexibility, it is clear that reliance on *ad hoc* allocations in the Latin America context has limited the productivity of certain linkages (Cardoso 1991; see also Box 5.7).

Box 5.14 Tensions in resource allocations resulting from insensitive application of government procedures: scaling up the treadle pump in Bangladesh

---

The treadle pump is a low-cost low-lift pump designed by the NGO Rangpur Dinajpur Rural Services (RDRS), which, with current sales of 65,000 per annum has made irrigation accessible to large numbers of low resource farmers who could not afford the mechanized tubewells which formed the mainstay of the government's programme.

Following a drought in the Rangpur and Dinajpur Districts of Bangladesh in 1989, the Ministry of Agriculture decided to order 100,000 pumps to be distributed free to farmers in the affected area. Standard tendering rules apply to orders of this size, and three companies which bid up to 35 per cent less than RDRS for manufacturing the pumps were awarded the contract. Checks imposed by the Tender Committee after over 13,000 pumps had been manufactured found them well below specification and so rejected them and cancelled the contract. The Ministry subsequently placed a much smaller contract with RDRS. Clearly, however, much wastage would have been prevented if a small amount of resources had been allocated to RDRS so that it could have monitored the quality of tender specifications, of bids and of early specimens of work by contractors.

Source: Orr *et al.* 1991; Hassanullah 1991

Critical issues in the development of linkages: thoughts for NGO and NARS managers

The majority of NGOs and NARS involved in the study recognized that linkages may generate positive results by allowing the comparative advantage of each side to be exploited in a synergistic fashion. But recognition of this potential is only a first step. As we have demonstrated, a range of mechanisms is needed to support the roles adopted by each side in linkage arrangements (cf. Agudelo and Kaimowitz 1991).

Recent institutional changes in the NARS of a number of Latin American countries have contributed to the establishment of structural linkages with NGOs. Formal operational linkages, however, are few in number. In Africa and Asia, by contrast, few structural linkages exist, but there is a comparatively high number of formal operational linkages with corresponding resource allocations. None the less, there is still overall a relative paucity of strongly-functioning linkage mechanisms, even where structural linkages exist. This suggests that much more thought needs to go into planning linkages. As an initial estimation, the following considerations seem to be especially important for such future planning:

- Before suggesting radical changes in the NARS the opportunities for developing links from existing procedures or institutional interaction should be explored.
- Linkage mechanisms do not function automatically. Once established they have to be adjusted to fit the needs and capabilities of the different actors involved. This is a difficult task which requires skilled intervention on the part of managers.
- Linkage mechanisms are not cost free, they require resources. Provision in terms of staff time and money should be planned.
- There are benefits in formalizing linkage mechanisms. A formal mechanism persists even when staff change. However, there are also risks: the flexibility which characterized the informal arrangement may be lost.
- With few exceptions the number and range of linkage mechanisms functioning at present is insufficient to allow close co-ordination between the two sides. If complementarities are to be exploited fully, the number and rang : of linkages needs to be substantially increased.
- Linkages depend heavily for their success on adequate mechanisms for joint planning and for the allocation of resources, in which numerous shortcomings are currently evident.
- Despite some evidence that workshops and published material have facilitated inter-institutional flows of information on relevant technology, there remains much scope for strengthening information flows from each side in forms which the other can readily use.
- Few formal mechanisms exist for evaluation in spite of its importance for detecting and correcting weaknesses in the co-ordination or development of linkages. When establishing links, managers should also devise appropriate ways of evaluating them jointly.
- The establishment of structural linkage mechanisms can improve the relationship between NARS and NGOs, and facilitate the establishment of a set of operational linkage mechanisms.
- However, the establishment of structural linkage mechanisms is not always possible. If the appropriate conditions do not exist it is better to establish specific operational linkage mechanisms where there is a clear interest in common.
- The establishment of structural mechanisms is not sufficient in itself to create functioning links. They must be modified as circumstances dictate and complemented with operational linkage mechanisms.
- Linkage mechanisms which are established by a central office may have little effect on field level staff. It is often best to begin locally when establishing or developing linkages. These can then respond to a more

clearly perceived need, for it is in the field and on-farm that gains from joint activities are most apparent.

GOVERNMENT-LED LINKAGES

The examples in the preceding section imply that most linkages have been initiated by NGOs. However, as we have suggested elsewhere, the state is often proactive in initiating linkages. While this may often be with the larger goal of using or controlling NGOs (Montgomery's (1988) 'bureaucratic populism') this is not always so. Here we give a brief review of several such government-led initiatives.

Five types of government-led interaction with NGOs are summarized in Table 5.3. These range from, at the one extreme, governments' efforts to set the overall policy and administrative framework in which NGOs operate to, at the other, project-specific initiatives taken by NARS.

Table 5.3 Types of government-led interaction with NGOs

Type	Examples	
1 Government sets the policy and administrative framework	Numerous countries	
2 Government facilitates NGO access to planning processes	Philippines	NGO desks
3 Government line Departments engage NGOs as part of a wider strategy	India:	Watershed development
	Philippines:	ERDB
4 NARS engage NGOs as part of their strategy, and may involve NGOs in NARS' planning process	India:	KVKs
	India:	NGOs and extension
	The Gambia:	FITT
	Bolivia:	CIAT
	Chile:	INDAP
5 NARS engage NGOs at a project-specific level	Senegal:	ISRA

Each of these types of government-led interaction is now considered in turn.

Type 1: government sets the policy and administrative framework

Government's role in setting the overall framework in which NGOs operate was discussed in Chapter 2. Briefly, it includes the types of NGO registration procedure evident in most countries, recent moves towards administrative decentralization in some countries which are likely to have implications for NGOs, and specific area-based co-ordination efforts led by GO staff. NGOs' room for manoeuvre, and the overall tenor of NGO–government relations is likely to be influenced also by government's policy orientation towards the rural poor, its capacity to implement pro-poor policies, and the extent of poverty-focus in ATD.

Type 2: government facilitates NGO access to planning processes

Again, elements of this type of government initiative were touched on at earlier points. For instance, enhanced NGO participation in local-level planning is one objective of proposals for administrative decentralization in Nepal (Box 2.3), and area-based initiatives by District Forestry Officers in Kenya allow NGOs the possibility of contributing to planning processes (Charles and Wellard 1993; Otieno 1992). However, none of these initiatives has addressed the central issue of the high transaction costs incurred by NGOs in attempting to inform themselves, and, where necessary, influence the higher-level planning processes of government. The only comprehensive initiative to do so encountered in the case studies is that taken by the Philippines government in creating 'NGO Desks' at several levels in the hierarchy of individual line departments (Box 5.15). These serve, on the one hand, to provide a first point of contact for NGOs' enquiries of government and, where necessary, to make government activities and staff attitudes more NGO-oriented.

Type 3: government line departments engage NGOs as part of a wider strategy

Two examples serve to illustrate this type of government–NGO interaction: first, as noted in Chapter 4 (Box 4.9) the Philippines Department of Environment and Natural Resources, as part of a strategy to re-establish government credibility in upland areas which had been neglected during the Marcos years, developed and disseminated a number of improved agri-livestock packages through local people's organizations (POs). As part of this process, government is helping to form the necessary organizational structures where POs do not currently exist or are deemed unsuitable.

Second, in India, growing concern over the increasing gap in living standards between irrigated and rainfed areas led to the establishment of a Rainfed Farming Systems Cell in 1987, which was given the mandate of pursuing sustainable farming systems approaches based on integrated water-shed development. A proposal is currently being considered by donors for funding collaboration with NGOs on a pilot scale in forty watersheds each of some 1,000 ha. It is envisaged that NGOs will take on the roles of:

- Creating awareness among rural populations of the need for watershed-based approaches;
- Providing training to junior government staff in participatory diagnosis;

- Conducting evaluations of the social impact of government-led interventions.

Box 5.15 'Outreach Desks' in the Philippines, and project-based NGO–GO collaboration with Southern Mindanao Agricultural Project (SMAP)

The NGO Outreach Desk in the Philippines Department of Agriculture (DA) was initiated in 1986, but strengthened in 1987 following the promulgation of a new National Constitution, Article II of which declared that 'the State shall encourage non-governmental organizations, community-based or sectoral organizations that promote the welfare of the nation'. Much of its early work sought to classify particular types of NGOs in ways acceptable to both sides, to establish a code of conduct for GO staff in dealing with them, to identify how the DA might establish working relations with them, and to conduct workshops among both NGO and GO staff with the objective of breaking down barriers between them.

The Outreach Desk in 1990 became responsible for the preparation of the five-year EC-funded SMAP which draws on the specialist skills of NGOs working in or near the proposed project area. These include:

- The Mindanao Baptist Rural Life Centre to train farmers, extension workers and other NGOs in the principles and application of Sloping Agricultural Land Technology (see Box 3.6);
- Resource and Ecology Foundation for the Regeneration of Mindanao (REFORM) to conduct socio-economic research for identification and planning of project activities;
- Santa Cruz Mission to identify sustainable ways of increasing the productivity of inland fisheries.

Since SMAP started in 1992, no conclusions can yet be drawn. However, a number of difficulties have emerged in the course of project preparation:

- The need for sensitive GO approaches to provide NGOs with a role wider than that of being merely an implementation arm of government;
- Tensions among some NGOs over acquisition of SMAP resources;
- NGOs' tendency to indulge in generalized policy advocacy instead of concentrating on the specifics of policy formulation.

Source: Fernandez and del Rosario 1993

Potential problems with the proposals include the complex decision-taking hierarchy which makes it unlikely that grassroots influence on strategy issues will be achieved, and the assumption that groups capable of representing the interests of the poor will emerge under the aegis of existing, frequently highly inequitable institutions such as *panchayats* (Seth and Axinn 1991; see also Box 5.16).

Type 4: NARS engage NGOs as part of their strategy

Several examples of government-led initiatives in NARS–NGO interaction were presented in Table 5.2. These included: the strategies of collaborating with NGOs in local-level trials and dissemination pursued by the Gambian agricultural research department under the Farmer Innovation and Technology Testing programme (Box 5.11); efforts by the Chilean NARS (INDAP) to develop and implement programmes of agricultural extension jointly with NGOs (Box 5.6), and the work of CIAT (Bolivia) in setting up a Department of Technology Transfer to implement an extension programme designed with and for 'intermediate users' of technology (Box 5.8 and 5.9). Furthermore, at CIAT NGOs are invited to participate in the annual research planning exercises during which CIAT's broad annual strategy is determined (Velez and Thiele 1991).

Box 5.16 NGO–government collaboration in watershed management in India

The State Watershed Development Cell (SWDC) of the state of Karnataka operates Watershed Development Programmes in nineteen of its districts, and seeks collaboration with at least one NGO in each. Its perception is that NGOs have a comparative advantage in identifying with the rural poor where and how interventions might be made, and in organizing groups from villages within the watershed to assist in construction works and in maintenance of the structures, once completed. The SWDC contacted a number of potential NGO partners through the Federation of Voluntary Organizations for Rural Development in Karnataka (FEVORD-K) and in March 1991 held a workshop for twenty of them in which the concepts were discussed and a number of issues raised for clarification (Bhat and Satish 1993). The SWDC's approach to watershed development is innovative both in the extent of consultation with rural people through NGOs, and in the creation of inter-departmental (i.e. agriculture, forestry, horticulture) umbrellas at district, division and state levels in order to co-ordinate planning and implementation. These proposals to work with NGOs in district WDPs draw partly on experience gained on a pilot basis with one NGO (Myrada) in Gulbarga District since 1984. This pilot exercise highlighted the need for strong co-ordinating mechanisms and clear division of responsibilities among line departments and between government and NGOs.

A further example of NGO–government collaboration in watershed management in India is found in the Task Force model developed in the Ministry of Agriculture of the national government to co-ordinate activities among departments in the planning and implementation of watershed management (Seth and Axinn 1991, summarized in Farrington and Lewis 1993). NGOs' roles are perceived as: creating awareness; providing training to staff of government and other NGOs in systems diagnostic approaches; and the evaluation of social impact. Major difficulties with the proposals – which form the basis of an aid request for $5.25 million – is that NGOs were not consulted about their perceived role, and that, since they are placed at the bottom of a strongly hierarchical structure, they are unlikely to be able to articulate grassroots concerns to decision-takers.

Source: Bhat and Satish 1993; Seth and Axinn 1991

Two further examples, both from India, illustrate other ways in which strategy-based links with NGOs are being developed.

First, the Indian Department of Agriculture and Co-operation has proposed a pilot scheme for five states in which extension would be contracted out to NGOs. NGOs' roles are expected to include the organizing of groups of farmers to visit demonstration sites and research sites, and the procurement and distribution of training literature. GOs will provide technical support to the NGOs and assist by supplying farm inputs through them (Farrington and Lewis 1993).

Second, despite the fact that in India the responsibility for agricultural extension lies with the states and not with the central government, the Indian Council for Agricultural Research (ICAR) operates a number of Farm Science Centres (Krishi Vignan Kendras – KVKs) intended to serve as centres for demonstration and training in 'scientific farming'. A number of these have been located at NGOs in recent years in an effort to strengthen the research capacity of the latter. However, the KVKs themselves have been small, understaffed and ill-equipped and have realized little of their potential for two-way interaction with higher-level research institutes under the ICAR (Satish and Dipankar Saha 1993; Satish and Kamal Kar 1993). It should be noted that in one case, however, the additional resources provided by a KVK have allowed a research organization established by and for medium/large scale tea planters to address the needs of small-scale tea-growers.[12]

Type 5: NARS engage NGOs at a project-specific level

As Tables 5.1 and 5.2 indicate, the great majority of interactions observed at project level between NGOs and NARS are NGO-initiated. However, a detailed example is provided from Senegal by Dugue (1993) of initiatives by the NARS (ISRA) to work with local membership organizations. For instance, ISRA's work with ARAF included the testing of (mainly) organic fertilizer technologies with ARAF's members, and training schemes both for farmers and for ARAF's technicians. ISRA found the high levels of participation of ARAF group members in identifying and implementing appropriate trials particularly beneficial. The collaboration offered by both ARAF members and its technical staff allowed a single research officer to conduct some fifty agronomic trials on farmers' fields in 1990, and to provide technical and economic monitoring for eight farms (Dugue 1993).

SEQUENTIAL LINKAGES: NGOs INNOVATE IN THE EXPECTATION OF 'SCALING UP' BY GOVERNMENT

In several of the case studies, NGOs have developed technological, methodological, organizational or institutional innovations during work with their own clients, which they have subsequently sought to replicate on a wider scale.

Much of this replication has been attempted through different forms of interaction with government (in the expectation that government might adopt the innovations more widely). The NGOs' motivation for engaging in the linkage mechanisms we have discussed is sometimes to affect such a wider change by influencing how government thinks and operates, as much as it is to improve the functioning of the ATS as it currently exists. In the section 'Linkage mechanisms supporting institutional roles', however, we focused on the functional dimension of the relationship (linkages to make technology paths more effective): in this section, we focus on this second dimension of the relationship – to have wider impact.

The theme of wider impact takes us to the so-called 'scaling-up' debate. Some of the early discussions of scaling up derived from Sheldon Annis' (1987) question, 'can small-scale development be large scale policy?' Annis, and institutions such as the Inter-American Foundation, were tested by the challenge of how to 'stretch' a very particular sort of development dollar (Morgan 1990): the dollar spent on grassroots development through local organizations. Many individual NGOs have also begun to take on this question internally: how to stretch their own dollars (Clark 1991; Gubbells, pers. comm. 1992; Edwards and Hulme 1992).

The simplest vision of scaling up was to do more of the same on a larger scale – replicating project experiences. This could be done by the same organization, or by trying to replicate the organization's innovations in the actions of other organizations. As we will see, there are many examples of this. However, some argue that replication may not address structural sources of poverty and marginality, and so ultimately the NGO should be concerned with pressing for the reform of programmes and policies in other organizations – primarily state organizations (Annis 1987; Bebbington 1991c).

If replication and reform are 'what' to do, the question is then 'how' to do them. As we noted earlier on, Edwards and Hulme (1992) have recently tried to synthesize actual experiences of NGOs engaged in scaling up strategies, and come to four main 'how tos'. These merit repeating: they are:

- Working with, and within, government structures to influence policy and systems;
- National and international lobbying and advocacy;
- Strengthening organizations of the poor;
- Operational expansion of the NGO itself.[13]

We might add two other means whereby an NGO may be able to scale up the impact of its innovation or approach. The first is to work through the market. This is a mechanism, however, that can only be used for innovations that can be turned into commodities. The second is for the NGO to give training to other organizations so that they replicate the work strategies of the NGO.[14]

The strategies are not mutually exclusive. For instance, training may be given to government institutions. Similarly, if working with government involves the NGO in doing considerable amounts of contracted work it can require operational expansion of the NGO. As we have noted, however, such operational expansion can create problems for the NGO of sustaining its internal coherence and identity, and of maintaining the quality of what it does.[15] These sorts of tensions, along with the more general (perhaps characteristically populist) fear that the minute an organization begins to grow it will lose the qualities that allowed it to do people-centred development in the first place, lead many NGOs to prefer to avoid significant operational expansion as a scaling up strategy, and consequently we will not treat it as a separate category.[16] Similarly, we discussed the strengthening of popular organizations and NGOs' advocacy activities in Chapter 4. In the remainder of this chapter we will therefore concentrate on: (1) scaling up through collaborative links with government (including training); (2) tensions that may emerge for NGOs that seek to do advocacy work at the same time as they collaborate with government; (3) scaling up through the market.

Edwards and Hulme's discussion is the result of a workshop specifically about scaling up. At that workshop certain points of agreement emerged regarding the strategy of working with government that bear out sentiments expressed in the workshops conducted during the course of the ODI study. In short, there seems to be some agreement that, given the rigidity of bur-eaucratic systems, high staff turnover and low motivation, such a strategy should be regarded as a long-term process. Furthermore, the prospects of success were enhanced if NGOs began to work with government from the outset, and so became familiar with its operating procedures. The most difficult point of the strategy is breaking through from the pilot stage, at which developments are heavily dependent on the NGO and perhaps one or two government staff, to wider replication. And once again, the strategy is not void of political contradictions: certain methodological innovations, such as those involving the mobilization and organization of the rural poor, are likely to encounter resistance in government for political reasons; and becoming too closely associated with short-lived governments can lead the NGO to be sidelined by successor governments.

As we noted, linkages with government can be motivated both by a desire to improve operational effectiveness and to exert a wider influence over governmental activities. Thus some of the case studies discussed earlier in this chapter also present cases of scaling up. For instance:

- The MBRLC in the Philippines (Box 3.6; 5.15) is a case of area-based collaboration between government departments and NGOs, with the former clearly scaling up the technologies developed by the latter;
- The RDRS treadle pump (Box 5.14) is a case of inappropriate balance in resource allocations, but clearly also represented a potentially important scaling up by government.

In other cases, the NARS has scaled up methodological and institutional innovations developed by NGOs;

- In Chile, the fifty-five new Centres for Adjustment and Transfer of Technology are being established by the NARS partly on the basis of methodological work done by NGOs, and so represent a scaling up of their efforts (Box 5.6);
- In India, thirty-four agricultural training institutes in Gujarat adopted and so scaled up the participatory approaches to technology testing and training developed by AKRSP (Box 5.12);
- In Kenya, the Forest Extension Services Division modified a number of its practices in collaboration with, or in response to, the work of several NGOs (Box 5.13);
- In Thailand, the extension services widely adopted the farmer-to-farmer dissemination methods developed by ATA (Box 5.3).

Other cases not so far mentioned in which *actual* scaling up of NGO technologies by government departments has occurred include:

- The action-research of LP3ES on farmer management of irrigation in Indonesia, which resulted in a government initiative to pursue farmer management on a much wider scale, bringing LP3ES into an expanded role of research and facilitation (Box 5.17);
- The research and development work on animal draught conducted by church-based NGOs in northern Ghana was taken up by a bilaterally-assisted government project, which also facilitated expansion of the technology by the manufacture of implements (Box 5.18).

In other cases, although the NGO intends its work to be scaled up there has as yet been little impact:

- The United Mission to Nepal (Thapa 1993) intends eventually to hand over its research on coppice reforestation to a government agency capable of pursuing it on a wider scale, and its nursery activities to farmers' associations and local groups;
- PRADAN's work (Box 5.21) with co-operatives of tanners was intended to demonstrate to government the type and depth of interaction with tanners that is necessary to facilitate the introduction of new technology;
- MYRADA's work on the development of participatory rapid appraisal methods (Fernandez and Mascarenhas 1993), and the workshops which it has generated (e.g. Mascarenhas *et al.* 1991) are clearly intended to stimulate the implementation of these methods on a wider scale by other NGOs and by government;
- Through its seed index and seed exchange programme, Auroville (Giordano *et al.* 1993) clearly intends that its innovative approaches to wasteland regeneration be scaled up, but contacts with public sector agencies (other

Box 5.17 The scaling up potential of work by Indonesia's Institute for Social and Economic Research, Education and Information (LP3ES) with small-scale irrigation systems

Formed by activist intellectuals in 1971, LP3ES conducts socio-economic research, education and information exchange in the pursuit of integral development of human resources. It works extensively with government, seeking to change its centralized top-down modes of operation into approaches which allow local people more say in their future.

Government-sponsored expansion of irrigation in the 1970s and early 1980s increasingly spread into areas which already had some irrigation infrastructure of their own. However, the top-down character of new irrigation design meant that farmers had little interest in maintaining it. At the government's request LP3ES began research on social and technical aspects of irrigation development in the mid-1980s, bringing in 'community organizers' on the Philippines model (Korten and Siy 1988) and experimenting with different forms of water users' associations.

Information collected during these three years of action research later proved invaluable when LP3ES was asked to assist the government in designing procedures for handing over to farmers the maintenance of all irrigation schemes under 500 ha (totalling some 1 million ha in all) to farmers. To work successfully with the government of Indonesia required the prior building up of trust between NGO and GO staff, NGO willingness to pursue incrementalist approaches, and measured amounts of opportunism, advocacy and constructive criticism.

Source: Bruns and Soelaiman 1993

than universities) have been few to date, the principal linkages having been with NGOs and local communities.

One of the simplest means by which an NGO can engage in scaling up its methodological (and other) innovations is through giving training to other institutions (Boxes 5.19 and 5.20), and in particular to the staff of the public sector. This is an increasingly common strategy, and Robert Chambers talks about the emergence of NGOs who have as their main speciality the training of others. Myrada, in India, is an example of an NGO which increasingly specializes in training participatory rural appraisal techniques, and we have already noted the example of the AKRSP (Box 5.12). Of course, the possibility of giving such training to state functionaries depends greatly on the disposition of the state. In certain contexts, this is such that it is simply impossible. In Chile, there was no question of NGOs giving training to the NARS under the Pinochet government. Now however, the presence of a civilian and elected government has created many more opportunities, and has allowed some NGOs to pursue a strategy that was always their long-term intention – namely to change the way the state operated (as in the case of GIA – Box 5.19).

Finally, a novel form of scaling up the combined impact of work conducted by NGOs, universities and public sector research institutes was facilitated by

---

Box 5.18 Government scales up animal traction technologies developed by NGOs in northern Ghana

---

The church-based agricultural development programmes in northern Ghana include agriculture stations (such as Langbensi – Kolbilla and Wellard 1993) established by the Presbyterian Church in the mid-1970s, the Tamale Archdiocese Agricultural Project established by the Catholic Church in the early 1970s (Millar 1991), and the inter-denominational co-ordinating body known as the Association of Church Development Projects (Alebikiya 1993).

A high proportion of church-based agriculture stations and development projects have had active and successful programmes of animal traction, comprising the selection and training of animals, development of implements and management practices for a range of tasks, and the training of farmers in their use. In response, the German-Ghanaian Agriculture Development Project initiated a number of bullock training programmes between the late 1970s and mid-1980s, and in 1979 opened an implement factory to produce ploughs, ridgers and cultivators. Although a broadly successful replication, one design of plough was found to be excessively heavy, and there was wide concern over the high levels of subsidy provided on these implements which NGOs found difficult to match.

Source: Amanor *et al.* 1993

IIRR in the production of an agroforestry kit (Gonsalves and Miclat-Teves 1993). The initiative drew together the agroforestry experience of this diverse category of actors in order to produce a guide primarily for the Integrated Social Forestry Programme of the Department of Environment and Natural Resources which plans to allocate 25 hectare plots of forest land to low-income farmers for a leasehold tenure of twenty-five years. The diverse experience in agroforestry was brought together in a one-week workshop in which individuals presented their experience so that it could be incorporated into a general format prepared by an NGO–GO steering group. Single sheet presentations of participants' experience, thematically arranged, were prepared in advance, and discussed and modified at the meeting. Illustrations for the material were prepared during the week itself, and by the end of the week a draft guide, complete with reference material and training handouts, had been prepared.

The kit has been widely used within the Philippines, and parts of it translated into local languages. On balance, the initiative succeeded in combining the diverse material of different institutions into an accessible needs-based format. It also demonstrates the brokerage role that a widely-respected NGO can play in facilitating focused and intensive interaction among the representatives of diverse types of institutions.

All these cases take us back to a more general point we made earlier in the book. Much of the logic behind the action and existence of many NGOs (here we might exclude the more opportunistic NGOs) is precisely to affect broad-

Box 5.19 GIA: An NGO's Influence over NARS and public institutions through training activities

The department known as the Area of Small Farmer Development Strategies was founded within GIA in 1981. This department saw itself as providing:

1 Support for small farmer survival strategies under the prevailing monetarist model of the dictatorship years;

2 Experience in the use of a methodology which would serve as a basis for defining, together with other research institutions, a national agricultural development policy for the small farmer sector once democracy was restored in Chile.

During this period, GIA adapted the concepts of appropriate technology, production systems, farm-level research and of the micro-region to the conditions of the Chilean small farmer. The intention was that these approaches should be applied on a wider scale than merely within GIA.

As one of their efforts to multiply the impact and social utility of this conceptual, methodological and practical work, GIA gave professional training courses. Although these did not constitute a direct influence on the policies of other organizations (much less those of the state), they did offer the possibility of influencing the staff of those institutions. Under the dictatorship, this training was only given to other NGOs, with the aim of strengthening their institutional capacities. Since 1985, GIA has hosted an annual course which runs for ten months (five days a month), for twenty-five participants from different NGOs.

Since the return to democratic rule, however, GIA has begun to give training courses to professional and technical staff from public sector institutions such as INDAP, INIA and CONAF (National Forestry Corporation) in such areas as diagnostic methodologies, and small farmer production systems, and farmer organization. It has also given training in these areas to the technical teams of the different technology transfer companies that implement technical assistance programmes on contract to the state.

Source: Sotomayor 1991

based social and policy change. They may not in fact do this – indeed, in Chapter 4, we joined others in criticizing them for paying insufficient attention to the policy context of their actions. None the less, the logic behind their existence is to change policy – to scale up their impact. Anything less, and they become gap-fillers, providing small subsidies to macroeconomic policies that otherwise continue unquestioned.

COLLABORATION AND CRITIQUE: INCOMPATIBLE STRATEGIES FOR SCALING UP?

In the bulk of this chapter we have looked at how collaborative relationships between NGOs and the NARS have operated. We have done so with a reason: to suggest that in some instances the strategy has considerable merit. Not only can it increase the effectiveness of the actions of the institutions involved – it can also, as we have shown in the preceding section, be an important means for NGOs to scale up their impacts.

Box 5.20 CLADES: a Latin American NGO consortium

Towards the end of the 1980s, many Latin American agroecological NGOs were seeking ways to achieve a greater impact in terms of sustainable rural development; CLADES was formed in 1988 with this object in mind. By 1991, CLADES proposed that, in order to achieve this aim, the following would be of major importance: (1) an improvement in collaboration between NGOs, governmental organizations, academic and international institutions; (2) a strengthening of the agroecological perspective in these institutions; (3) a change in relations between NGOs and other institutions in Latin America and the international development agencies; (4) a strengthening of NGOs' institutional, technical and professional capacities; and (5) a division of labour between the different institutions according to their comparative advantages.

On beginning this undertaking, one of the main activities was to synthesize and diffuse information on agroecological innovations and work that was held in the many different and often isolated institutions in Latin America and (especially) the USA. Subsequently, this information was used for professional training activities organized and co-ordinated by CLADES. This training, although in large part aimed at NGOs, was also open to technicians from the state sector. In organizing these training activities CLADES' aims have been to foster co-ordination and communication between different types of institutions, and to strengthen the professional capacity of NGOs, above all in their use of an agroecological perspective.

At the same time, CLADES has organized seminars for NGO executives to discuss among themselves how best to respond at an institutional level to the politico-economic changes that are pressing NGOs to take a more prominent role in development.

Source: Altieri and Yurjevic 1991

Yet elsewhere, we have suggested that NGOs can use these same linkages as means of exercising pressure over public sector institutions, to urge them to reorient their policies. Furthermore, in Chapters 2 and 4 we also stressed the importance of NGOs adopting the role of policy critic – and indeed the need for them to take questions of policy change more seriously than they have done in the past.

These are different strategies of scaling up: the one collaborative, the other with an element of critique and at times conflict. Can a single NGO combine these strategies?

An NGO's collaborative activities with a public sector organization are likely to be strained if the NGO begins to become a more vocal critic of its partner. Yet there are cases in the research material that show NGOs combining these two strategies. How far the roles can be combined by a single NGO ultimately depends a great deal on the style and attitude of the people involved, and the relationship between them: tact and friendship are not rigorous theoretical concepts, but they can have a great bearing on such interactions. In CIAT (Bolivia), for instance, NGOs have been collaborators with, but at times quite severe critics of CIAT – but the personal friendships underlying these relationships allowed the collaborations to continue.

The possibility of combining such strategies is enhanced when the criticism is voiced in a more informal way to high level officials in government. Of course, informal pressure gives the NGO no means of sanctionable or vote-casting influence over public decisions: but it can be effective. There are several cases of this occurring in the case studies. Such a strategy, however, requires that the NGO staff have such contacts in the first place. A series of factors come into play in increasing the likelihood of this being the case: these include family ties,[17] shared political affiliations, or shared university backgrounds. Such contacts are also more likely where there has been some movement of staff between state and NGOs: in cases such as Chile, where many NGO people returned to government after the end of the Pinochet dictatorship, this is quite common.

A more frequent scenario in which critique and collaboration are combined is where some NGOs (who place particular value on links with government) specialize in collaboration, and others in critique[18] – though of course, such division of strategies is not usually a conscious decision.

For donors who are concerned that such critical pressure be brought to bear, then there is an important case to be made for supporting research oriented NGOs who are particularly likely to pursue this more advocacy-based form of scaling up.[19] However, their critique will be more informed, incisive and above all more legitimate if they also engage in some action work, and if their links to other operational NGOs are good. Also, for their critique to have an impact, they must have more powerful means of communicating it than a mere publication. Once again, informal contacts at a high level are crucial in this regard – and here donors can also help in establishing some such contacts, as for instance the informal NGO–government working groups that the Swiss Technical Co-operation has supported in Bolivia.

LINKAGES BETWEEN NGOs AND THE PRIVATE COMMERCIAL SECTOR

The fact that much mainstream discussion of research and extension has focused on public sector institutions has not only understated the importance of NGOs. It has also under-emphasized the role of the private commercial sector (Pray and Echeverría 1990). It is true that private commercial sector representation is not strong in the complex, diverse and risk-prone farming areas with which we are principally concerned, mainly because of poor communications with urban areas and the fragmented nature of markets.

However, in several countries the commercial sector is responsible for developing and distributing new technologies such as seeds and agrochemicals, and its outlets can often be found in the most remote of villages (Kaimowitz 1991).[20] Private interests also innovate in processing and market development in ways which impinge on low-income farmers. Finally, it is worth noting that the commercial sector is not only composed of multinational capital: it also embraces the many small-scale commercial artisans who engage in informal experimentation for the design of small farm tools and equipment.

Evidence is accumulating on the varying ways in which NGOs interact with some of this private sector activity. Several examples are found in the case study material, and in wider NGO literature. This is significant, for while many NGOs may have been critical of market 'penetration' and capitalist development in the past, their increasing interaction with commercial interests suggests that their approach is changing away from a simple critique of capitalism towards efforts to increase the capacity of NGOs themselves and of the rural poor to use the marketplace as a resource. Our analysis of case study material in the remainder of this section suggests that it may serve as a resource in two ways:

- NGOs may use the market as part of a strategy to scale up their innovations;
- Local groups may be encouraged to link more closely with the private commercial sector in order to enhance their own institutional sustainability.

The market and NGO 'scaling up' strategies

The use of the market to aid the dissemination of an innovation can be indirect (by developing a market for products that derive from that innovation) or direct (by distributing the innovation itself through the market). The experience of MCC in Bangladesh provides an example of the more indirect use of the market. MCC recognized that no matter how good its research on varieties and production techniques, soya would not 'take off' unless a market for this hitherto unknown crop could be found. As part of a market research effort, it therefore began to identify snack foods that could be produced from soya. MCC engaged the interest of a number of private entrepreneurs in this process, several of whom have now begun to market soya products and, in doing so, have helped to create a demand to which producers outside MCC's clientèle may respond (Buckland and Graham 1993).

Also in Bangladesh, RDRS used the market more directly to broaden the dissemination of its technology. The NGO engaged a number of private sector blacksmiths to produce the treadle pump as a commercial proposition. Much more was involved than simply a contractual relationship: RDRS's technicians worked with the blacksmiths to devise simplified solutions to production engineering problems compatible with adequate product quality (Orr *et al.* 1991). Quality quickly fell to unacceptable levels when, in an effort to 'scale-up' the technology, manufacturers were contracted by government through a crude tendering process from which the technical training link between NGO and manufacturer was absent (Box 5.14).

Other examples of NGOs' links with the private commercial sector in developing technologies include the widely-quoted work in the Arusha area of Tanzania by Lutheran World Relief in developing a manually-operated press to extract oil from sunflower seed. The designs were discussed with local blacksmiths, who provided ideas on how production might be simplified without sacrificing quality (Hyman 1992).

Also in Tanzania, the Mbeya Oxenization Project operated by the Mennonite Central Committee successfully worked with a local implement manufacturer to produce improved ploughs after attempts to work with parastatal manufacturers had failed. MCC's philosophy in a related area, the manufacture of axles for animal-drawn carts, has been similar: with its assistance, local commercial workshops have progressed from the manufacture of components to the production of complete axles, and subsidies on the finished product have gradually been reduced in an effort to make these efforts commercially sustainable (Graham 1992).

Linkages with the market as a means of strengthening grassroots organizations

In Chapter 4 we discussed the difficulties faced by NGOs in trying to determine at what point grassroots membership organizations had become strong enough for the NGO to withdraw, and in designing withdrawal strategies. Of the small number of case studies that have given serious thought to this issue, some introduced the notion of payment by the rural poor for the services that the NGO itself initially provided, but which were subsequently provided by its local partner groups. Thus, in Bangladesh, for instance, the local women working as paravets in BRAC's poultry initiatives (Box 3.9), in FIVDB's duck production (Nahas 1993) and in Proshika's livestock work (Khan et al. 1993) are paid by the recipients of their services; in India, part of the costs of local extensionists working in AKRSP's participatory research and training programmes (Box 5.12) are covered from within the communities themselves; and in CESA's efforts to improve the quality of seed potatoes in Ecuador, the costs of seed and agrochemicals provided by CESA for the first year's production on a community seed plot were recouped from the seed harvested in the second year (CESA 1991).

NGOs' support for tree nurseries falls in the same category: UMN in Nepal plans to encourage local groups to establish tree nurseries incorporating the species and nursery techniques which it has been developing as its own nurseries wind down – thereby creating new income-generating opportunities (Thapa 1993). CARE's work with women's groups in Kenya strengthens the economic opportunities offered by their existing nurseries (Buck 1993).

In a similar vein, a number of the projects allow local groups to link in new ways into markets providing inputs in support of new technologies. Thus, the Mutuoko Agricultural Development Project in Zimbabwe encouraged farmers to organize bulk purchases of fertilizer for improved maize, which were both cheaper than individual supplies and were delivered close to their fields (Mvududu 1993).

However, some NGOs' efforts go further than this, seeking to enhance their local partner organizations' ability to manage new technology, to draw on the government services they require, and to negotiate with the private commercial sector. Two examples can be cited: in the first, NGO-support for landless irrigator groups in Bangladesh (Mustafa *et al.* 1993; Wood and Palmer-Jones 1990) was premised on their ability to learn how to use and maintain pumps, and to obtain credit for government, but also to sell water to farmers. More fundamentally, this support derives from major NGOs' perception that the intensification of agriculture in Bangladesh offers opportunities to the rural poor for which reliable access to land is not a prerequisite. As such opportunities expand, economic power is spread increasingly across those holding assets other than land. The imperative perceived by a number of NGOs in Bangladesh, therefore, is to focus less on issues of land reform and more on the identification of ways in which the rural poor – particularly the landless – can participate in emerging opportunities independent of access to land.

A second example is provided by PRADAN's work with a low-caste group of flayers in India (Box 5.21) which stimulated the 'ownership' of technology in the same way, together with links both with the market and with government, but encouraged the local group to draw lessons from its interaction with the market which, in turn, could enhance the efficiency of its use of the new technology.

Underlying these NGO initiatives towards cost recovery and the identification and exploitation of new commercial opportunities is the conviction that the long-term sustainability of innovations in which the rural poor engage is less well served by autarky than by the development of a capacity to interact with the wider economy, in cost-recovery, in the efficient acquisition of inputs and in the sale of the goods and services generated.

CONCLUSIONS

The title of this chapter gave us a menu of phrases to choose from for our description of NGOs interactions with the state: have they been reluctant partnerships, hostile confrontations or productive synergies?

Most rhetoric from NGOs might lead one to believe that there was in fact little relationship at all with the state – and if there was, any interactions were often hostile confrontations and at best reluctant partnerships. In the course of the research, we have found this rhetoric to be still quite common among NGOs.

While there is evidence to support this popular wisdom, we have suggested that recent changes in the broader political and economic context of many countries at least offer the scope for more productive interaction between these

Box 5.21 PRADAN: scaling down technologies, links with the private commercial sector and the formation of self-sustaining groups

Professional Assistance for Development Action (PRADAN) is an Indian NGO established by a number of natural and social scientists in 1983 with the initial objective of placing its professionals in local NGOs for several months at a time in order to address constraints they faced in technology and social organization, and in interacting with government agencies. This approach has now been replaced by one in which Development Support Teams comprising PRADAN staff service local groups and assist local NGOs to interface with GOs on an area basis. A further objective is to draw socially concerned young professionals to work in villages, so providing a learning space in this 'field university' and, through links with their former universities, stimulating curricular change.

Vasimalai (1993) gives examples of PRADAN's work in de-mystifying and scaling down technologies developed by government for use by the rural poor in the production of mushrooms and raw silk, and in processing hides and skins. The last of these is described here.

Traditional methods of curing leather by low-caste groups rely on salting and drying, resulting in a low-value product. Government efforts to develop and promote improved (chrome leather) tanning are vested in numerous institutes, but their success is limited by poor outreach capacity, wide dispersal of traditional tanners, the high cost of improved technologies and the need for effective links with the private commercial sector in input procurement, marketing and quality control if new technologies are to be adopted on a sustained basis.

PRADAN acted as a bridge between an existing tanners' co-operative, government departments and the private commercial sector to facilitate introduction of the new technology. It first introduced the idea to the co-operative, working with it for six months to gain its confidence and to allow the idea of new technology to be 'owned' by the co-operative. It then helped to arrange grants and loans, arranged training for co-operative staff in the relevant government institute, designed a processing plant scaled down to one-third of its conventional size, worked with co-operative members and the commercial sector to construct the plant, and provided a leather technician to assist in its initial operation. A literate member of the co-operative noted the treatment details of each batch of leather and compared them with the quality of the product in an effort to keep down the costs of achieving a given quality level.

PRADAN withdrew in April 1990, after three years of support to the co-operative. The outcome has not simply been the introduction of new technology, but empowerment of the co-operative to control the technological process, and to interact with government credit and advisory services, and with the private commercial sector in obtaining feedback on quality.

At the same time, PRADAN's interaction with government in this 'pilot project' has demonstrated, first, the depth of interaction that government research, training and support institutes must seek with local groups if they are to 'own' the technology and, second, the co-ordination needed across training, input supply, plant design, credit and commercial marketing if the potential of improved tanning by local groups of artisans is to be realized.

Source: Vasimalai 1993

different organizations. We discussed some of these changes at the beginning of the chapter.

Furthermore, there is in fact much evidence of very diverse and often productive forms of interaction between NGOs and the state, which have in some instances been fully collaborative and had positive outcomes for all involved.

Identifying these interactions is one thing: finding pattern in them is quite another. Much of the chapter has tried to find this pattern, adapting a terminology developed for the analysis of linkages between institutions in the public sector. It should be emphasized though that the pattern we claim to have found is very much a retrospective one: very few NGOs or NARS have thought systematically about their linkages with each other. Indeed, this very lack of *ex ante* planning is a major source of weakness in these relationships.

A wide range of linkage mechanisms, both formal and informal, were identified. Many of them had specific resources allocated to them. We drew particular attention to the significance of structural linkage mechanisms (as opposed to operational mechanisms), for these begin to allow NGOs a say in the decision making and resource allocation of the NARS (and vice versa). These, however, are still quite infrequent, and comprised: joint committees for the management of research, joint annual planning meetings, and representation by one side on the Board of the other. The wide range of operational linkage mechanisms included joint activity along the sequence of stages in research, including problem diagnosis, research design, technology testing, evaluation, dissemination and training.

In the remainder of the chapter, a number of looser interactions between NGOs and government were discussed. These included innovation in technology, methods or social organization by NGOs with the intention that government might replicate them on a wider scale. Brief consideration was also given to NGOs' interaction both with their clients and with government in an advocacy role, and to the types of expectation embodied in government-led initiatives to interact with NGOs. While some of these are designed to meet NGOs' requirements (such as the NGO Desks located in Philippine line departments) others (including the Indian proposals to bring NGOs into extension and into watershed management) appear to be based more on the requirements of government than NGOs.

The existence of these linkages leads to several questions, one of the most interesting of which is why NGOs sustain a confrontational rhetoric about the state when some of their interactions are far more collaborative in vein, and productive in results. Clearly institutional identity has a large say in this. Many NGOs, and certainly most in this study, are more at ease with the role of being critics of government, and of identifying their own (supposed) strengths against a backdrop of their claims about government inefficiency, corruption and bias. There may be an element of truth in all those claims about parts of government – but clearly NGO practice suggests that other parts of government are not necessarily such a lost cause. It is probably time for NGOs to have a finer grained analysis of the state and its different institutions. It is also time for them to begin reappraising their identities.

A second question is whether there is pattern in how these linkages have been constructed. We delay a full discussion of this theme until the final chapter, for it takes us back to consider the conceptual points introduced in Chapter 2.

However, understanding the factors that have given rise to more successful linkages is an important input to any future planning of these mechanisms. Here it is important to reiterate that linkages do not function by themselves. Whether they are structural or operational, the mechanisms underpinning linkages have to be tailored to the specific characteristics of the institutions concerned, and to the opportunities and constraints posed by specific types of interaction. Very rarely has this been done in the past. Consequently, successful linkage is more likely if linkages are planned according to the institutions' needs and capacities, and are then given the necessary resources to sustain them.

Finally, we are faced with the challenge of how to assess these linkages. On the one hand we can assess them in terms of the attitudes the participants have towards them – in other words, in terms of the actors' interpretations of NGO–NARS linkages. These interpretations range widely. Some interactions, particularly of the advocacy kind, are still seen as confrontational. Most interactions though, are seen in a more ambiguous light. While certain positive outcomes are recognized by the people involved, many weigh these gains against what they perceive as the drawbacks of these linkages: that they imply a loss of time, a loss of autonomy and in some cases (for NGOs in particular) an undermining of their identity. Of all of these, it is the loss of institutional autonomy and independence that perhaps inevitably results from linkage, particularly what we call structural linkage, that most vexes the organizations involved – again perhaps particularly the NGOs.

For these reasons, there is still much scepticism about linkages. NGOs and NARS may be moving towards partnership – but reluctantly so.

On the other hand, we can also assess these interactions in terms of their impacts on the livelihoods of the rural poor – who in large measure are indifferent as to whether GSOs lose or retain their autonomy. Assessing this impact is difficult, but the evidence gathered together in this chapter suggests that the gains from linkages can be very significant – and while it may not always be the poorest who benefit, the people who benefit are still poor by any standards.

To call these fruitful impacts the result of a synergy between NGOs and NARS might be pushing the case a bit too far – but it seems clear that, if socio-political and institutional obstacles to interaction are overcome, there is much complementarity between the tasks that NGOs and NARS do well.

NOTES

1 Thanks to Simon Zadek for this turn of phrase.
2 Assertions which we too must admit to having made on occasions.
3 The analytical framework for this section, particularly as applied in Tables 5.1. and 5.2 draws extensively on the work of Graham Thiele in Bebbington and Thiele (1993). We are grateful for permission to use it here.
4 This analysis refers back to the discussion of the ATS in Chapter 2.
5 Although in itself complex enough, Thiele's analysis is deliberately simplified by postulating that technology development follows a linear pathway from basic to adaptive stages, whereas Biggs stresses that 'pathways', where they can be identified, depend on particular configurations of institutional conditions prevailing at any point in time, and so are unlikely to be linear.
6 Where useful distinctions between applied and adaptive research *can* be made, they are made in relation to specific case studies, not as a component of the wider analysis.
7 This is the category on which the ISNAR studies focused (Merrill-Sands and Kaimowitz 1991).
8 A third African case – FITT in the Gambia – involved dissemination by NGOs as an integral part of NGO–NARS collaboration in adaptive research (Box 5.11).
9 It is also worth noting that FUNDAEC's sister NGO, CELATER, is organizing courses to familiarize public sector staff with the results of this study's research on NGOs.
10 It should be noted that the degree of order and planning implied by our use of these terms and categories does not imply that the organizations involved gave the same degree of prior thought and planning to the linkages. In some respects, we are making sense out of chaos, or applying structures that in reality do not exist. However, such structuring is a necessary first step for any attempts to plan these linkages more systematically in the future.
11 The classification developed by Merrill-Sands and Kaimowitz (1991) identified planning and review processes and the development of communication devices as separate categories. For simplicity, they have been grouped under 'joint professional activities' here.
12 At the tea research station of the United Planters' Association of Southern India (UPASI – see Swaminathan and Satish 1993).
13 In the end Edwards and Hulme (1992) conclude that the most appropriate strategy depends on context (once again!) and the characteristics of the particular NGO.
14 Perhaps the most remarkable impact of the 'scaling up by training approach' has been the recent spread of participatory rural appraisal (PRA), in large measure as a result of a barrage of training inspired, or done, by the International Institute of Environment and Development and the Institute of Development Studies (see, for instance, Mascarenhas *et al.* 1991).
15 See the discussion in Chapter 2 of the contrasting experiences of the Chilean NGOs, AGRARIA and GIA.
16 It is the case that some NGOs, particularly northern NGOs, have grown rapidly in recent years. For many south-based NGOs, however, funding limitations are a great obstacle to such a strategy.
17 On one occasion, one of the authors was in a meeting with the (conservative) Director of an NGO who, part way through the meeting called his brother to arrange for the author to discuss the same theme with his (social democrat) brother. His brother directed a national programme of rural development.
18 This was found, for instance, in the Philippines (Farrington and Lewis 1993).

19 For instance, CEDLA in Bolivia, GIA in Chile, CEPLAES in Ecuador, CELATER in Colombia.

20 In some cases, the necessary research is conducted in-country, in others it is imported from elsewhere.

6

WHERE FROM, WHERE AT, WHERE NEXT?

Books evolve in the process of writing them, and as the book changes, so too does the title. 'Between the State and the Rural Poor' was one of the earlier working titles for this book – and although the title was ultimately assigned its place in the waste-paper basket, it still captures the awkward position occupied by NGOs. While they generally profess a closer affinity to the poor than to the state, they bear more resemblance to the state than they do to the poor – and in most of their activities they operate in a manner that is more akin to the state than to any organization of the rural poor.

In the end we have said more about the uneasy relationship between NGOs and the state than we did about their relationship to rural people – hence the change of title to *Reluctant Partners*, a summary, it seemed to us, of the ambivalent feelings NGOs and governments have towards each other. None the less, the uneasiness of the middle ground that NGOs occupy has loomed large in this book.

The development and dissemination of agricultural technology (ATD in our shorthand) served as our entry point for analysing this uneasy relationship. In line with more systemic approaches, we kept our conceptualization of agricultural technology broad – our definition of agriculture encompassed on-farm, off-farm and watershed level interactions in the management of crops, trees, animals and fish; we used the term technology to embrace hardware, management practice and knowledge systems.

Despite this broad definition, the focus on ATD gave us a set of tangible issues on which to hang our analysis of inter-institutional relationships. At the same time, the focus on ATD had its own justification. Agriculture remains of central importance to the livelihoods of the rural poor; hence interventions in agriculture are a central part of government and donor policies of poverty alleviation and rural development. However, the most appropriate means of organizing these interventions continues to be debated in the light of the poor performance of governments in providing agricultural research and extension support to the rural poor – a deficiency which contrasts with the measure of, albeit limited, success that some NGOs have achieved.

THE CASE STUDIES IN CONTEXT

This study of NGOs began in 1989 – seventy case studies later, what can we say we have learned? Perhaps the main lesson has been that it is difficult to make any general statement about them at all. NGOs are a very diverse bunch, and the ways in which they engage with the state, the rural poor and agricultural technology is in turn dependent on the local and national contexts within which they operate – and of course, these contexts are also very diverse.

In a sense, we have tried to control for some of this diversity by focusing the case studies on specific types of NGO. We have concentrated on the larger NGOs, and those which have some significant experience as regards their relationship with the state – be this a collaborative or a more conflictive and critical interaction. As a consequence, little attention has been paid to the numerous small NGOs that exist in the countries in which research was conducted, nor to those many NGOs whose work rarely brings them into contact with government. Most of these smaller NGOs are unable to bear the transaction costs involved in identifying the appropriate government departments and individuals with whom to interact, and in coping with the inflexibilities and caprices of government bureaucracy (Wellard and Copestake 1993). On the other hand, the larger NGOs may be acting as brokers for the smaller ones with which they collaborate, or as pioneers in making government easier for them to access directly, and so reducing transaction costs in the longer term.

To continue stressing context dependence and diversity may not appear very constructive to those who want 'relevant' research, yet we would counter that this is not so. 'Relevance' is not only a characteristic of research that says *how to act*; research is equally relevant when it points to the questions that must be asked *before acting* in order to increase the likelihood that the final outcome of the action will be coherent with the initial intentions. It is on that criterion of relevance, and on the basis of our restricted sample of cases that we venture the tentative general statements in the remainder of this chapter.

WHERE FROM: THE STATE AND NGOs IN CONTEXT

In much of the policy making and programme design of recent decades, it has typically been assumed that the state and its agencies were the appropriate vehicle for implementation of policies and projects. In part, this assumption grows out of Keynesian inspired and import-substitution models of development, which cast the state a central role in strengthening domestic production capacity. In some countries, the role also derived from a vision of the state as principal protagonist in a process of nation building and self-determined development in the post-colonial period.

The 1980s were not a happy decade for these development models. With increasing vigour, commentators have drawn attention to the inefficiencies in resource allocations, the discrimination against rural areas, and the imbalances between public sector and private commercial development that grew out of this heavy state involvement in production, marketing and development planning. The crisis of public sector finances, particularly in Latin America and Africa, has only compounded this disillusion with the performance of public sector agencies. At the same time, the growing donor concern with 'good government' has also fed more vocal criticism of inefficient, inequitable and authoritarian tendencies in state bureaucracies, highlighting how biased resource allocations by the state stem not merely from poor decisions, but from structures which prevent many groups in society from exercising influence over government's decisions, and from holding it accountable for its actions.

The general dissatisfaction with state intervention deriving from these observations has been compounded by a more specific dissatisfaction with the state's role in agricultural development.[1] Whilst different governments' policies towards agriculture and the rural sector have been driven by varying combinations of production and equity criteria,[2] the agricultural development models pursued to achieve these diverse objectives have been based almost universally on the 'modernization' of agriculture through the use of improved genetic material and of energy-intensive external inputs such as agrochemicals and mechanized tillage and pumping. These models have been premised on the view that traditional practices can be rapidly *transformed* into modern agriculture. Their success, however, depends on a range of factors which have often been lacking – the availability of adequate infrastructure for input supply and marketing being but one of these.

At the same time the success of the approach requires the presence of new institutional structures that will provide the necessary support to farmers. Public research and extension services are one such institutional structure, and one whose capacity to promote the 'modernization' of small farmer production has been questioned with increasing cogency. Numerous studies have highlighted the requirement that research should not only be client-oriented, but should also work with clients in the diagnosis of problems, and in the screening, testing, evaluation and dissemination of technical solutions. They have also stressed the need for strong feedback to researchers of farmers' views on the technologies under trial. It seems increasingly the case that government research and extension services (NARS) are faced by severe obstacles in meeting these challenges. Some of these obstacles are characteristic of wider problems in government and include political interference to use research services for patronage, bureaucratic inertia and inflexibility, budgetary restrictions and delays, inadequate operational budgets, hierarchical and centralized management structures, and the absence of performance-related reward systems. Other problems are more specifically related to the characteristics of research and extension, and include inadequate scientific management, disproportionately heavy reductions in budgets for field operations, institutional

179

barriers to communication and to joint activities between research and extension, the inadequate incorporation of clients' needs into priority setting in research programmes, and inappropriate methods for research and dissemination.

As public sector budgets are reduced, the capacity of NARS to confront these challenges is further undermined. This is particularly so in those demographically and economically 'small' developing countries whose NARS have few resources to cope with high levels of agroecological and socio-economic diversity. For these NARS it is difficult to acquire 'lumpy' research infrastructure, such as laboratories and libraries, or to support the critical minimum mass of effort required for conventional types of investigation requiring replication and multi-site testing.

Concern over the intractable nature of many of these constraints has led to growing interest in the contribution that NGOs can make to ATD. In particular, interest is clearly growing in the extent to which there are practical complementarities between NGOs and government, and to which NGO approaches might be incorporated into governmental programmes.

Much of this interest in NGOs has been fuelled by a body of suppositions (what we might call 'NGO-lore') about their qualities. This lore asserts that NGOs operate with a 'people-centred' approach to agricultural development in which the type and pace of change is determined by NGOs' clients.

The conventional view is that NGOs' strengths lie in their: (1) participatory approaches to needs assessment; (2) holistic, issue-oriented approaches, which allow constraints to be addressed not only across the range of agricultural activities, but also in support services and in such wider contexts as education, health and nutrition; (3) focus on livelihood enhancement among the rural poor; and (4) concern with exercising a 'demand-pull' on government services, where possible through the strengthening of local groups. Furthermore, it is also assumed that they work with low external input, low risk, environmentally benign technologies that are developed and disseminated in such a way as to reinforce indigenous knowledge systems and local institutions.

Against these strengths are contrasted a range of NGO limitations. These weaknesses include a limited capacity for research, weak links with wider policy or scientific spheres, inability to acquire (or share) lumpy assets, weak interaction with other NGOs conducting similar work, the small, localized nature of their efforts; and, in a wider vein, the absence of mechanisms to ensure the representativeness of NGOs and their accountability to the rural poor.

These strengths and weaknesses are an almost perfect mirror image of those attributed to NARS, leading to claims that collaboration with the NARS would build on institutional complementarities which could benefit NGOs, NARS, and not least the small farmer.

In large measure, this study has argued that there is considerable truth in this NGO-lore, although there is also much variability in how far NGOs live up to the claims made for them. Furthermore, in practice, two sets of factors

180

place obstacles in the path of moves towards complementary NGO–state interaction. First, there is great diversity among NGOs as regards their philosophies, objectives and mode and scale of operation. Governments are similarly diverse in the development objectives they pursue, the role they envisage for the rural poor in future scenarios, the means by which such roles are to be achieved, and their capacities for implementing planned interventions.

Second, each NGO has its own social history and was created for specific purposes. Some of these purposes may relate narrowly to the performance of specific functions, but, in other cases, NGOs were, for instance, established as a form of opposing corrupt or repressive governments, and are therefore circumspect about rushing into association with public institutions – even under more democratic administrations. Many governments are equally reluctant to engage in collaboration with NGOs. Non-democratic regimes may simply see them as potentially subversive, and democratic governments may view their actions in social development as implied criticism of the state's shortcomings. In addition, some governments resent what they see as the 'diversion' of external funds into NGO activities.

WHERE AT: NGOs AND AGRICULTURAL TECHNOLOGY DEVELOPMENT

While NGO-lore might lead us to think that all NGOs operate with low input, agroecological perspectives on agricultural resource management, this is an over-simplification of the truth. While the scheme used in this book was also a simplification, we identified two main tendencies among the technological foci of the case study NGOs. Those approaches that worked with some variant of fossil fuel based technology and high yielding varieties we denominated as *production-oriented*, and those which emphasized low external input agriculture we called *agroecological* approaches.

Within each tendency there was further variation. Among those 'production-oriented' NGOs, some used high external input technological packages in a largely uncritical fashion, simply assuming that they constitute an improvement on indigenous production systems. Others select this technological option after analysing what would be (to their mind) the most effective response to local conditions: these approaches are much more sensitive to existing institutional, organizational and agricultural contexts. This more pragmatic selection of technology is paralleled among the 'agroecological' NGOs, many of whom promote low input options on the grounds that they constitute the most appropriate option for the context within which their farmer clients operate. El Ceibo's promotion of organic cocoa as a high value export product is the most evident example of this. Finally, among these 'agroecological' NGOs are those that promote low input agriculture because they (like the purist modernizers) simply believe it to be inherently the 'correct' technological option. This is most apparent among those NGOs whose actions are driven by ideological perspectives on organic agriculture, the environment or ethno-development.

Reflecting this diversity of technological perspective, there was also diversity in the extent to which NGOs drew on indigenous knowledge in establishing what broad types of innovation would be acceptable to their clients. Overall, though, the more important sources on which NGOs drew for innovation lay outside local communities and, in many cases were located in modern scientific knowledge spheres. Many NGOs sought to strengthen local capabilities to absorb, experiment with and modify such innovations. The emphasis was on strengthening indigenous management capacities rather than indigenous technologies *per se*.

Whatever the technological focus with which the NGO works, some characteristics of their work were more likely to be held in common than others. These included a concern for developing participatory methods for ATD, and social organizational forms for the sustainable implementation of technology. Also common was a tendency to have a more 'issue'-oriented focus in which barriers between research, dissemination and feedback are minimized, and in which constraints within and beyond agriculture are addressed. Particularly notable in some contexts of high population density are NGOs' efforts to develop technologies tailored to the needs of women and the landless.

Sensitivity to local context varies among NGOs, but strong field presence gives some a higher awareness of local conditions than found in government, and this determines how technological change is implemented in practice. Thus, grassroots-sensitive efforts to 'modernize' peasant farming draw-down technologies relevant to local needs and seek to devise support systems compatible with local socio-economic, cultural and institutional conditions. The impact of NGOs' work in ATD was more difficult to assess. There is evidence suggesting that some innovations devised or promoted by NGOs have had wide economic impact. However, comparisons of costs and benefits could be made only in a few cases, and only in one case was it possible to draw comparisons between the cost-effectiveness of NARS and of NGOs. Evidence in these areas is likely to remain fragmentary since:

- Most NGO efforts are exploratory in nature and are intended to produce largely qualitative benefits, such as a strengthened capacity for experimentation among farmers;
- NGOs' and NARS' objectives do not coincide exactly: NGOs' objectives tend to be broader and less easily quantified;
- Monitoring and evaluation capacity is under-resourced on both sides.

However, most of the case study NGOs were seeking impacts that were technically, economically and institutionally sustainable. Consequently, a recurrent theme in their work was the establishment of strong local institutions capable of enhancing the technical capabilities of the rural poor and strengthening their position *vis-à-vis* the public sector, other interests in society and (in many cases) the market.

The wide range of innovative features of NGOs' work in ATD should not be taken to imply that they are about to have a major impact on the livelihoods of the rural poor – their efforts remain too small, fragmented, and poorly co-ordinated. Perhaps the most significant implications of their experience lie in the lessons they generate that have potential for being scaled up by governments.

WHERE AT: NGOs AND OTHER ACTORS

NGOs and the rural poor

Many of the current proposals for NGO-government interaction that are emanating from governments and donors cast NGOs in a service delivery role. In these versions, the relationship of the NGO to the rural poor is that of provider to receiver. Most of the NGOs considered in this study have a somewhat different perspective on this relationship. While they may be delivering services, they see their role as facilitators not providers: working with the rural poor to strengthen the capacity of grassroots organizations to identify their needs and meet them from their own resources, or by demanding a response from the state. At the same time, many of the NGOs saw themselves as innovators, developing new approaches to poverty alleviation, including natural resources-related technologies and management practices, for 'scaling up' by the state.

Claims about NGOs' ability to engage in participatory, grassroots organizing activities with the rural poor have, on the whole, been uncritically speculative.[3] The material in this study demonstrates considerable diversity in the nature and quality of NGOs' relationships with the rural poor. In particular it is clear that the label 'participation' is applied to a wide range of interactions. 'Participation' in NGO projects can be differentiated according to *depth* of participation – this can range from relationships in which the NGO designs and then provides services itself, through to more 'empowering' approaches. At the same time, they can be differentiated in terms of the *breadth* of subject matter focus – some NGOs work with the poor on a single technology, others on a whole range of issues relevant to local livelihood development.

The two dimensions are often interrelated, in that NGOs using more empowering approaches are inevitably led into a wide thematic focus, since the livelihood constraints identified through processes of conscientization are rarely specific to a single sector such as agriculture. Paradoxically, the wide scope of action identified in these processes can overstretch the NGO, whose technical and organizational abilities to facilitate improvement in a wide range of issues are inevitably limited, and thus lead sometimes to disillusion among clients. GOs, by contrast, generally have a narrow sectoral mandate, which makes them disinclined to enter the types of deeper dialogue with clients that are likely to bring cross-sectoral issues onto the agenda. Nevertheless, within the narrow context of identifying, developing and delivering adoptable technologies, there remains enormous scope for GOs' performance to be enhanced – and their poverty reach to be improved – by better use of the project preparation, monitoring and feedback tools currently at their disposal, and of participatory methods which could be incorporated from NGOs' experiences.

At the same time, while we must be realistic about what can and cannot be achieved in the public sector, and the NARS in particular, we must also recognize that participation by itself will not solve rural poverty. Indeed, this study endorses the general tenor of two other empirical studies which have suggested that while NGOs' main strength is in promoting participatory development and strengthening local capacity to manage development, the principal weakness is that they have difficulty in coming to grips with, and then addressing, the processes and relationships underlying rural poverty. As Carroll (1992) and Riddell and Robinson (1993) have shown, few NGO income-generating projects manage to reach the poorest of the poor, many promote solutions that are not financially sustainable, and few NGOs are concerned with the creation of employment opportunities. A critical weakness is that much NGO work is conducted in isolation from wider policy issues – yet if they are to have a broad-based positive impact on poverty, they must ultimately change this policy environment. It may, then, be the case that too much NGO attention has been paid to the qualitative aspects of development, such as participation, and not enough to the harder questions of income generation. For as long as economic survival is the prime concern of the rural poor, they will be unable to participate fully in, for instance, the projects and local organizations that NGOs (and others) are using as building blocks for self-managed development and for enhanced grassroots pressure over the state. Seen in this light, economic improvement may therefore be both a prerequisite for, as well as a result of, broad participation by the poor in the development process.

Our analysis of the empirical material on NGOs' relations with clients also highlights two other sets of issues. The first of these pertain to NGOs' relationships with the poor. Many NGOs have no formal mechanisms of accountability to their clients. Aside from being somewhat contradictory with their commitment to participatory democracy, this lack of internal participation in NGO decision making provides governments – especially those whose policies are criticized by NGOs – with ready excuses for not working with them. In a somewhat similar vein, although the question of phasing out the support they have provided to local membership groups is prominent in many NGOs' rhetoric, in practice, it is fraught with difficulty in anything other than the least complex interactions. In general, NGOs rhetorical commitment to self-managed, sustainable and participatory development exceeds the reality of their practice.

A further set of issues pertains to the implications that NGOs' stance *vis-à-vis* the poor have for their relationships with government. NGOs will face potential difficulty in collaborating with government if at the same time they are acting as advocates on behalf of the poor. Some NGOs committed to one of the two approaches have therefore decided to forgo the other, or to take it forward through links with other NGOs. Also, empowerment has implications not only for NGOs' relations with the poor, but also for the creation of new relations between the poor and government. Thus there is a case for arguing that, at the same time as they work in trying to strengthen the organizations of the rural poor, NGOs ought also to seek means of making government more accessible to those organizations.

NGO–government interaction in ATD

A variety of preconditions are likely to favour collaboration between NGOs and government in ATD. At a macro-political level, there needs to be some convergence on overall development objectives – a convergence which will be more difficult to achieve if there is a history of conflictive relationships between repressive governments and NGOs. More specifically there needs to be convergence in their visions of the future of the rural poor, and then of the technological and organizational strategies that are most likely to achieve those objectives.

However, while convergence facilitates collaboration, it is not necessary that there be congruence in development objectives before states and NGOs work together. For instance, while more profound levels of participation and empowerment may be central concerns for many NGOs, this has not prevented some of them from seeking to identify how less radical change in the government sector might be promoted in order to generate positive outcomes for the rural poor.

Much of Chapter 5 was dedicated to analysing the roles that have been adopted by NGOs and government research and extension services when they have found it possible to interact (whether on an unstructured or a managed basis) in this broadly collaborative fashion. Cases were identified in which NGOs disseminated technology developed by NARS (and vice versa), and in which NGOs conducted research with NARS and then disseminated the results of that research – these were was the most common types of of interactions.

Some of these interactions have not required contacts between the organizations involved – as when NGOs work independently with technologies generated by NARS. In other cases, however, there is direct contact. The relationship mediating these direct contacts we called 'linkage mechanisms'. These have been both formal and informal, and many had specific resources allocated to them.

A wide range of relationships were observed. In broad terms they fell into two broad categories: those *structural* linkage mechanisms in which one organization has an influence over resource allocation and programming decisions in the other organization; and *operational* linkage mechanisms in which organizations collaborate around more specific project activities. While certain *structural* linkage mechanisms were observed, these occurred only in a few cases, reflecting the fact that it is more complicated for one organization to begin allowing another organization a formalized influence over its internal decision making processes. None the less, there is evidence, particularly in cases where the NARS is undergoing a restructuring process in which closer links to NGOs are being sought, of the following forms of inter-organizational relationship: joint NGO–NARS committees for the management of research, joint annual planning meetings, and representation by one side on the board of the other. There was evidence of a far wider range of *operational* linkage mechanisms, including joint activity along the sequence of activities in research, from problem diagnosis, through research design, technology testing and evaluation, to dissemination and training.

The evidence suggests that linkages do not function by themselves. Whether structural or operational, if linkages are to be effective, the mechanisms underpinning them have to be tailored to the specific characteristics of the institutions concerned, and have to be managed to meet the opportunities and constraints arising both from changes in socio-political context and from the evolution of specific types of interaction. This sort of careful planning, however, happens only rarely.

Both NGOs and governments have initiated linkages, though in doing so they have often had different objectives. For NGOs, linkages have the dual concern of gaining access to research expertise and technological resources in government, and of trying to scale up NGO-generated innovations through the government apparatus. This scaling up has also been sought through NGOs giving training to government staff and through more conflictive critical interactions, in which the NGO lobbies government to change its form of operating or its policy position.

For governments there are similarly a variety of reasons for initiating linkages with NGOs. In some cases, government institutions have sought to 'use' NGOs for the implementation of public policy, or to find means of monitoring and co-ordinating NGO activity. In other cases, however, the motivation has been to make government more accessible to NGOs (such as the NGO Desks located in Philippine line departments).

Perhaps inevitably, NGOs have found it easier to interact with the newer, smaller government departments having mandates embracing, for instance, agroforestry or the environment. This is partly attributable to NGOs' experience (and, often, government's inexperience) in these subject areas, so that the government is keen to learn from the NGO. Another factor, however, is that in these newer departments attitudes tend to be less entrenched, and views of what constitutes 'research' less rigid than, for instance, in departments of agricultural research where considerable status is attached by scientists to their own specialist skills. Such scientists sometimes have difficulty in appreciating the role of the (often) social science-based skills and experience

There is also some, albeit limited, evidence of government efforts to 'scale up' initiatives taken by NGOs within the public sector. In some cases, this has involved government in incorporating into its own work participatory or systems-oriented methods developed by NGOs. In fewer cases, governments have tried to work with technical innovations developed or initially used by NGOs; this has been the case, for instance, with technologies such as crop varieties, hillside management strategies, animal draught power, and mechanical technologies such as the treadle pump. In far fewer cases has government tried to work with the social organizational innovations developed by NGOs for the management of certain technologies. However, there are cases where this has occurred, for instance with government adopting NGO methods for passing on the management of irrigation systems to users' associations in Indonesia.

Evidently, there are many hurdles that must be passed before government successfully scales up NGO-generated technologies. Aside from the challenge of identifying an appropriate technology, and a sustainable system for managing it a local level, adequate channels of communication, training and support must be maintained between NGOs and government. Furthermore, it is not only NGO–state interaction that must be effectively organized – the internal structure and procedures of the government institution itself must also be such that the organization is capable of sustaining the delivery of appropriate forms of support to farmers.

NGOs and commercial agents

The case studies were primarily about the ways in which NGOs and government have interacted. However, in the course of preparing the studies it has also become apparent that in several cases, NGOs have also sought means of engaging with market agents for the sustained development and dissemination of technologies. Some of these links have been between the NGO itself and the market agents, and in other cases the NGO has promoted direct contacts between the rural poor and the market agent. These examples are interesting and merit more analysis than the little attention they have been given in studies of NGOs hitherto. Taken in their context, these have the potential to be strategies for enhancing the capacity of the poor in negotiating their market relationships, and for finding mechanisms of supporting local innovations that are more sustainable to the extent that they do not depend on institutional interventions. The limits and potentials of these strategies should be the topic of further research.

Donors, NGOs and governments

It has been suggested that NGOs' principal sources of funds explicitly or implicitly influence their selection of activities and modes of operation. This is a hypothesis that has strong intuitive appeal, and for which crude support can be found in the field. For instance, those donors who support market development and the withdrawal of the state are directly implicated in the creation of what we have termed 'opportunist' NGOs. These NGOs are quite ready to play their part in a model of society in which service provision is sub-contracted to the private sector and in which the social contract is rewritten to one in which the state becomes merely an enabler rather than a provider. There seems little doubt that this direct interaction between funding agencies and opportunist NGOs is set to spread as structural adjustment continues to bite, and service delivery institutions are needed to fill what would otherwise be a void in rural areas (and as increasing numbers of former civil servants seek new employment opportunities).

However, to suggest that donors are determinative in the last instance would merely be to rewrite a crude dependency theory for the 1990s. Nor would it do justice to the fact that there is much diversity in the ways that donors behave *vis-à-vis* NGOs. There are experiences in which the role of the funding agency has been one of *facilitation*, supporting NGOs as they pursue their own agendas. In other cases (probably the majority of NGO funding until recently) the donor has played a role of *constructive influence*, exercising some leverage over the design of a programme. In other cases donors have certainly *determined* project design, at times in a way that *distorts*, working against the better instincts of the NGOs – as, for example, when CIDA placed pressure on Bangladeshi NGOs (and in the end unsuccessfully) to play a training role in the government's co-operative programme for the poor.

Similarly, there is a range of ideological persuasions among donors. Indeed, there is almost as much diversity among funding agencies as among NGOs. While the dominant political philosophies in such major donors as USAID are well-known, other bilateral donors are driven by different considerations. Some Scandinavian countries, for instance, have long prioritized aid for participatory-type development projects to a small number of countries having left-of-centre governments. Furthermore, many NGOs rely to a high degree on north-based NGOs or international foundations for their funding. Here, also, there exists much diversity: some agencies are more closely linked to government than others, some to particular religious views and so on. The important point is, that with the recent explosion of interest in NGOs, South-based NGOs in many contexts are to an extent in a buyer's market: they are somewhat able to choose a funding agency whose views closely reflect their own – which is quite different from suggesting that these views are imposed on them.

It is important to recognize that NGOs are able to manoeuvre within this diversity as they pursue their own strategies. It is not difficult for NGOs to identify those donors who are most likely to support the particular type of proposal the NGO wants funding. While in general donors are less inclined to support more radical NGOs, there are, as those NGOs know, many ways to repackage a radical programme.

The capacity of southern-based NGOs to exercise choice over their funding agencies can also be enhanced by charismatic leadership. In such cases, the relationship between an NGO's actions and its source of funds is certainly not direct: both are driven by an additional factor – the quality of leadership.

In certain cases, perhaps particularly those where leadership is charismatic, NGOs may even influence the actions and thinking of their donors. Indeed, to influence donor thinking is one of the main objectives of the Latin American NGO, CLADES. While it is unlikely that a single NGO will influence the thinking of a whole donor institution, it is also not the case that all decisions within a funding agency reflect the supposed 'institutional' position. Even the apparently direct relationship between the headquarters of North-based NGOs and their own branches in the South is, in practice, often characterized by a wide diversity of views and actions. Ultimately, individuals, departments and country desks make decisions, and at this level NGOs are more able to exercise influence.

The relationship between South-based NGOs and their funding agencies is therefore complex. While an analysis of NGOs' sources of funding may therefore provide insights into their choice of project activity, the relationship should not be interpreted as causal, and evidence which is at least as illuminating is likely to be gleaned from analysis of the activities that NGOs undertake – or avoid.

However, having said this, we must recognize that donors exercise influence. Consequently, their funding strategies will have significant sway over the direction of NGO–government interaction in the future. The most powerful tendency here is currently towards government retrenchment and instrumental interactions in which NGOs are used (via sub-contracts or otherwise) to implement public policy. While this *may* improve the efficiency of some services, we would suggest that care needs to be taken to avoid drawing the more catalytic NGOs into such contracts, to the extent that their innovative capacity is undermined. At the same time, donors could use their position to encourage contacts between governments and NGOs to help build mutual trust between the two. Projects involving inter-institutional staff secondments, NGO contracting of NARS research, and informal working groups could contribute much to a process in which these different actors become more aware of each other's needs, constraints and modes of operation. As the research has suggested, the existence of personal networks cutting across institutional boundaries is one of the most important prerequisites for successful collaborations – supporting staff secondments and working groups would help lay the bases of such networks.

WHERE NEXT? THEORY AND THE FUTURE

Gazing into the future is rarely advisable for researchers, especially if they fear being proved wrong. Perhaps rashly, then, we close the book with a few reflections on possible future tendencies that might emerge in NGOs' work in rural areas with the poor, as well as in their links with government.

In the first two chapters of the book we outlined a conceptual framework that would underlie our analysis of agricultural technology and the inter-institutional relationships underlying its development. In this framework, the process of ATD was located within a wider socio-political and economic context which we argued set many of the conditions within which individual actors operated (even if ultimately that same context is the result of human agency). While NGOs (and others) might have the potential to influence the policy making process in that wider context, for most (if not all) it remains the context within which they must conceive their operations and strategies.

However, we insisted that while that context may structure actions, it does not determine them, and individual actors have a certain room to manoeuvre. How they take advantage of that flexibility will, among other things, be an effect of their own histories, strategies and social networks. Consequently any analysis of inter-institutional relationships should pay attention to how actors have operated within socio-economic and political structures in the past. These two considerations of, what in the broadest terms might be considered structure and agency (cf. Giddens 1979), will influence the extent to which apparent functional complementarities between different organizations are exploited.

Although this framework has been largely implicit in the empirical chapters, it has proved helpful in understanding the case material. It is clear, for instance, that certain political environments have precluded NGO–state collaboration. Indeed, NGOs and states have often had quite different development objectives – they could not have agreed on the 'why' of collaboration let alone the 'how'. Similarly, in the contemporary political economic context(s) of structural adjustment(s) there is much disagreement over objectives – even though the contemporary situation in many countries is now in several respects far more conducive to collaboration, the shadow of structural adjustment still constitutes an obstacle to closer relationships between NGOs and the state.

Yet, within these constraints there has been much scope to negotiate collaboration. Much of this scope has been engineered by actors drawing on social networks that cut across institutional boundaries. It is also the case that the diverse opinions among NGOs, and the diversity within the state's own structure has meant that even when, say, a Chilean NGO disagreed fundamentally with the dictatorial politics of the military government, it could find, and then work with, individuals within the state, however informal this link may have been. The political economic context limited what could be achieved through these interactions, but it was not determinate: it was still possible to find means of improving the quality of the work of all involved. This negotiation through the soft spots of harshly drawn institutional lines has allowed people to take advantage of institutional complementarities – in the above noted Chilean case, NGOs could take advantage of the NARS' technical expertise, and those individuals in the NARS who wanted to work with small farmers but were constrained in doing so by their institutional mandate, were able to sidestep this constraint, by working with the NGOs – even if this had to be on a Sunday.

Given, then, that this framework has helped our analysis of past and present, it seems appropriate to use it as our optic through which to look to the future. Again, though, we reiterate a caution we made in Chapter 2: we are not trying to generalize about the explanation of future realities, but simply about the sorts of question that should be asked of them, and the issues that might be expected to impinge on them.

In that spirit, while making a generalization about the future socio-political contexts of Africa, Asia and Latin America is absurd, we can perhaps point to tendencies that are more likely to occur than not. Following the structure of the book, we will comment on the implications for agricultural technology, for NGOs' relationships with the poor, and for NGO–government interactions.

NGOs and agricultural technology

To the extent that structural adjustment policies take hold, and remove import subsidies, trade barriers, and overvalued exchange rates, then South-based agriculture will be under pressure to compete with imported foods, and to develop capacity in agroexports of both indigenous and non-traditional products. If, at the same time, the market imperfections subsidizing high external input, fossil fuel based technological packages are removed, such packages will become more expensive and less competitive, unless used very efficiently (Kaimowitz 1991).

For NGOs working with small farmers, such changes will present acute challenges. Certainly there will be much more pressure to identify and work with lower input, but productive technological options, in order to help farmers survive increasing competitive pressures. While this may mean there will be more scope for working with agroecological options, this will have to be a very hard-nosed agroecology – one less inspired by ideological conviction, than by a recognition of the need to increase farm efficiency and income. This imperative of efficiency will also mean that technical assistance to farmers will have to go much further than it has in the past in imparting the skills of numeracy, market negotiation, financial resource management and farm accounting.

At the same time, to the extent that these policies aggravate rural poverty, vulnerability, and perhaps rural–urban migration, there will be growing pressure to work with technologies that increase family level food security. To this end, low input technology, storage and perhaps most significantly, techniques for urban gardening and peri-urban agriculture will require NGOs' attention.

NGOs and the rural poor

These policy environments will have diverse implications for NGOs' relationships with the rural poor. Most critically they will probably make it yet harder to identify economically viable income-generating strategies. This demand, alongside the 'imperatives of efficiency' noted above, will place pressure on NGOs to increase their professional competence in economics, market analysis, and farm management. Here, however, like research institutes in the UK, they may find it difficult to pay salaries that are high enough to attract the professional competence they will need.

The future of the relationship between NGOs and the organizations of the rural poor, and thus indirectly of NGOs' broader work in participatory and empowering approaches, is very uncertain, and will undoubtedly exhibit much geographical variation. Looking back at the crisis years of the 1980s in Latin America, some commentators see one of their main heritages in the disarticulation of popular organizations (Jenny Pearce pers. comm.). Such organizations have had great difficulty in surviving inflationary pressures, and identifying viable strategies in such adverse environments. At the same time, to the extent that the growing poverty of their members has made the costs of participating in any sort of activity more burdensome, these organizations have had difficulty sustaining the interest of their membership. The potential power of 'grassroots movements for global change'[4] is therefore questionable in such difficult contexts. In these circumstances, rather than being under increasing pressure to pass more project management onto the rural poor themselves, NGOs will be required to continue supporting, nurturing and keeping alive fragile organizations.

And yet, while the 1980s were a lost decade for some, in other cases, popular organizations have gained greater strength (Bebbington et al. 1993). In these cases, it will be more appropriate for NGOs to begin adopting the role of a consultant service to these organizations, so that the organization can draw on them (and pay them) when they need particular forms of support. How common such instances may become, however, is open to doubt: while in countries like Ecuador this may be a possibility (Bebbington et al. 1993), in West Africa the rural poor seem far less likely to be able to exercise demands on anyone, government or NGO (Gubbels 1992).

Another implication for this relationship with the rural poor derives from the sorts of skill the NGO imparts, and the sorts of popular organization it promotes. Much of the popular education and group formation work that NGOs have done, has been oriented toward strengthening the poor's assertiveness and capacity to make demands on the state for services. Although we may not like to admit it, to the extent that state services are increasingly diminished, such skills and organizations may begin to become obsolescent: there may be little state left on which to make claims. While there may be other institutions on which to make claims, it remains the case that as the state withdraws, much more development initiatives will, somehow, have to be self-managed and ultimately self-financed. This will require different sorts of skill and different forms of local organization, perhaps based on clearly identified economic strategies.

NGOs and their relationships with government

We have suggested that the current policies fostering public sector reform and more transparency in government are, on balance, likely to lead to more frequent and potentially fruitful NGO–government interaction. Before developing that theme, however, we will point to several major caveats.

We have already pointed to the difficulty NGOs face in deciding whether or not by working with a retrenching public sector they are endorsing and supporting that retrenchment. This uncertainty will hinder collaboration with government, and will have the effect that some NGOs will continue to maintain their independence.

Perhaps a more worrying concern for NGOs, however, is how far the neo-liberal economic policies that go hand in hand with public sector reform may undermine any gathering processes of democratization. The impact of adjustment has, for instance, led to poverty-related disturbances in several countries – Venezuela's so-called IMF riots being one example. As these disturbances continue, threatening more serious political instability, the threat increases that a hard line, perhaps military, government may return to power. The various coup attempts in Venezuela in recent years are just one example of this. Similarly, as different forces nurture ethnically and regionally based movements whose activities have often led to conflicts with other groups, elites who control the means of violence in countries may feel the need to revert to repressive government in order to contain such conflict.

There have been democratic springs before and they have seen no summer. Left-of-centre NGOs and popular organizations cannot help but be aware of this and worry how far a collaboration with government now will make them more vulnerable to future repression should that government change.

These points aside, it seems to us that other forces will continue to create space for NGO–state interaction. Although interest groups within government may try to undermine public sector cutbacks, it seems inevitable that these government reforms will continue.[5] As this happens, and even if development and welfare expenditure is reduced, government and donors will look for alternative channels for implementing and funding social and economic development work at the grassroots. In some cases, we can expect a growing tendency toward co-financed NGO–government ATD projects. In many other cases, governments will try to use NGOs to implement public programmes. Some existing NGOs will therefore grow, and will as a result have difficulty in sustaining their institutional coherence and identity. Some of the NGOs in our study may well go in this direction. The strengths and innovative capacities of some may be lost.

In many other cases, new NGOs will continue to emerge. They will do so partly in order to take advantage of the laws favouring NGOs: thus one can expect the label of 'NGO' to be abused by business people who will create 'NGOs' in order to be allowed tax-free imports, for instance.[6] Other NGOs will be created to capture the donor funds earmarked for NGOs, and the sub-contracting opportunities with government.

These abuses may well lead to a growing popular dissatisfaction with NGOs, and the label will begin to lose some of the legitimacy it has had in previous years. At the same time, the proliferation of NGOs will cause mounting problems of institutional chaos, to which government will ultimately have to respond by identifying and promoting mechanisms of co-ordination – the few noted in Chapter 5 are likely precursors to a far larger range of mechanisms in the future. In a context of decentralizing government administration, it is likely that these mechanisms will be stronger at sub-national levels. NGOs may also respond to this proliferation, through forming networks that co-ordinate activities, and that at the same time are intended to mark out the members of the network as 'legitimate' NGOs, as distinguished from the opportunistic variety. How far those networks will become the contact point with the state will depend on whether they can resolve their own problems of internal co-ordination and transparency.

Similarly, this increasing use of NGOs by the state will probably require much more careful thought on the part of all involved as to how to organize linkage mechanisms. This care will be forced by early failures and hiccups in NGO–government collaborations, and will also be forced by the likelihood that intentions to work with NGOs will feature increasingly in national plans. The dominantly *ad hoc* approach of the past will thus slowly change. This can only be a good thing.

While the state may try to take the initiative in promoting co-ordination mechanisms and linkages, it will not be able to set the agenda independently. As it becomes more dependent on NGOs for the implementation of pro-grammes, its own bargaining strength will be reduced, and that of NGOs correspondingly increased. NGOs will therefore have more scope to press for policy changes – though probably within the limits set by the international financial institutions.

Whether NGOs attempt to take advantage of this scope to negotiate by interacting in collaboration with government will depend on how they assess the wider situation, and the potential returns to them of closer interaction and collaboration. In turn this response will be influenced by the history, background and strategy of the NGO and the key decision makers within it, as well as the resources, size and felt needs of the NGO. It will also depend greatly on the conditions in the public sector. In cases where state institutions are exceedingly weak, and become (in the short term at least) yet more inefficient as a consequence of cutbacks, then NGOs may well decide there is little left in the state on which they can draw, or which can be improved through the exercise of pressure. As the agricultural sectors of some economies become almost entirely dependent on international finance institutions and the ministry of finance, NGOs may also decide that there is little room to influence policy. Sadly, they may be right.

Ultimately, and this takes us back again to the problem of diversity, different NGOs will respond in different ways – and different countries will offer different contexts and differing amounts of room within which to manoeuvre. Thus among countries, there will continue to be some which enjoy more NGO–state interaction than others: within countries, different NGOs will pursue their own specific forms of interaction with government – some will interact in collaborative implementation activities, others in collaborative technological and policy research, others in more confrontational modes. Depending on their histories, objectives and strategies, some will be more, and others less, reluctant to enter into partnerships with government.

As donors try to plan for NGO–state interaction they should pay heed to this diversity. Just as they are learning that there can be no blueprints for implementing structural adjustment and good government programmes, so there can be no blueprints for coping with their consequences.

NOTES

1 The theme of the 1992 World Bank Agricultural Symposium was 'Public and Private Roles in Agricultural Development.' Champions of the public cause were not obviously in a majority.
2 Policies in Thailand, for instance, have been determined largely by a perceived need to penetrate export markets. Conversely, in India, although production efficiency is given a high priority, it is accompanied by a wide range of measures, both within and beyond agriculture, geared towards poverty alleviation.
3 There have been times when we too have made such claims.
4 The term is Paul Ekins' (Ekins 1992).
5 At the time of writing the Nepalese civil service had just been radically pruned, the government making redundant all those staff over a given age, and who had served more than a certain period.
6 This is already a recognized problem in some countries – an example is Sierra Leone.

REFERENCES

Abed, F.H. (1991) 'Extension services of NGOs: the approach of BRAC' paper to National Seminar on GO–NGO Collaboration in Agricultural Research and Extension, 4 August, Dhaka.

Adriance, J. (1992) 'User friendly agroecology networks', *Grassroots Development* 16(1): 44–5.

Agudelo, L.A. and Kaimowitz, D. (1991) 'Institutional linkages for different types of agricultural technologies: rice in the eastern plains of Colombia'. *World Development* 19(6): 697–703.

Aguirre, F. and Namdar-Irani, M. (1992) 'Complementarities and tensions in NGO–State relations in agricultural development: the trajectory of AGRARIA (Chile)', *Agricultural Research and Extension Network Paper No. 32*, London: Overseas Development Institute.

Alebikiya, M. (1993) 'The Association of Church Development Projects (ACDEP) in northern Ghana,' in K. Wellard, and J.G. Copestake (eds) *NGOs and the State in Africa: Rethinking Roles in Sustainable Agricultural Development*, London: Routledge.

Altieri, M.A. (1987) *Agroecology: the Scientific Basis of Alternative Agriculture*, Boulder, Colorado: Westview Press.

—— (1990) 'Agroecology and rural development in Latin America'. in M. Altieri and S.B. Hecht (eds) *Agroecology and Small Farm Development*, Boston, Massachusetts: CRC Press, pp. 113–20.

Altieri M.A. and S.B. Hecht (eds) (1990) *Agroecology and Small Farm Development*, Boston, Massachusetts: CRC Press.

Altieri, M and Yurjevic, A. (1991) 'Influencing north–south and inter-institutional relations in agricultural research and technology transfer in Latin America: the case of CLADES', paper presented at 'Taller Regional para América del Sur: Generación y Transferencia de Tecnología Agropecuaria; el Papel de las ONGs y el Sector Público', 2–7 December 1991, Santa Cruz, Bolivia.

Amanor, K., Wellard, K., and Denkabe, A. (1993) 'Ghana: country overview' in K. Wellard, and J.G. Copestake (eds) *NGOs and the State in Africa: Rethinking Roles in Sustainable Agricultural Development*, London: Routledge.

Annis, S. (1987) 'Can small-scale development be a large-scale policy? The case of Latin America'. *World Development* 15, supplement: 129–34.

Annis, S. (1988) 'Can small-scale development be large-scale policy?', in S. Annis and P. Hakim (eds) *Direct to the Poor: Grassroots Development in Latin America*, Boulder and London: Lynne Reinner, pp. 209–18.

Annis, S and Hakim, P. (eds) (1988) *Direct to the Poor: Grassroots Development in Latin America*, Boulder and London: Lynne Reinner.

REFERENCES

Arbab, F. (1988) *Non-governmental Organizations: Report of a Learning Project*, Cali, Colombia: FUNDAEC.

Ayers, A.J. (1992) 'Conflicts or complementarities? The state and NGOs in the colonization zones of San Julian and Berlin, eastern Bolivia', *Agricultural Research and Extension Network Paper No. 37*. London: Overseas Development Institute.

Barsky, O. (1990) *Políticas Agrarias en América Latina*, Santiago, Chile: CEDECO.

Bates, R.H. (1981) *Markets and States in Tropical Africa: the Political Basis of Agricultural Policies*, London: University of California Press.

Bebbington, A.J. (1990) 'Farmer knowledge, institutional resources and sustainable agricultural strategies. A case study from the eastern slopes of the Peruvian Andes', *Bulletin of Latin American Research* 9(2): 203–28.

—— (1991a) 'Sharecropping agricultural development: the potential for GSO–government co-operation', *Grassroots Development* 15(2): 20–30.

—— (1991b) 'Indigenous agricultural knowledge, human interests and critical analysis: reflections on the role of farmer organizations in Ecuador', *Agriculture and Human Values* 8, (1/2): 14–24.

—— (1991c) 'Planning rural development in local organizations in the Andes. What role for regional and national scaling up?' *RRA Notes*, No. 11: 71–4, IIED, London.

—— (1992) *Searching for an Indigenous Agricultural Development: Indian Organizations and NGOs in the Central Andes of Ecuador*, Working Paper No. 45. Cambridge: Centre of Latin American Studies.

—— (1993) 'Sustainable livelihood development in the Andes? Local institutions and regional resource use in Ecuador', *Development Policy Review* 11 (1): 5–30.

Bebbington, A.J. and Carney, J. (1990) 'Geographers in the international agricultural research centers. Theoretical and practical considerations,' *Annals of the Association of American Geographers* 80(1): 34–48.

Bebbington, A.J. and Farrington, J. (1992) 'The scope for NGO–government interactions in agricultural technology development: an international overview', *Agricultural Research Extension Network Paper No. 33*, London: Overseas Development Institute.

—— (1993) 'Governments, NGOs and agricultural development: perspectives on changing inter-organisational relationships', *Journal of Development Studies* (2): 199–219.

Bebbington, A.J. and Thiele, G. (1993) *NGOs and the State in Latin America: Rethinking Roles in Sustainable Agricultural Development*, London: Routledge.

Bebbington, A., Carrasco, H., Peralbo, L., Ramón, G., Torres, V.H., and Trujilli, J. (1993) 'Fragile lands, fragile organisations. Indian organisations and the politics of sustainability in Ecuador', *Transactions of the Institute of British Geographers* 18 (2).

Bernstein, H. (1977) 'Notes on capital and peasantry'. *Review of African Political Economy*, 10: 60–73.

Bhat, K.V. and Satish, S. (1993) 'NGO links with the Karnataka State Watershed Development Cell: MYRADA and the PIDOW Project', in J. Farrington and D. Lewis (eds) *NGOs and the State in Asia: Rethinking Roles in Sustainable Agricultural Development*, London: Routledge.

Biggs, S.D. (1989) 'A multiple source of innovation model of agricultural research and technology promotion'. *Agricultural Research and Extension Network Paper No. 6*, London: Overseas Development Institute. (Republished in shorter form in: *World Development* (1990) 18 (11): 1481–99.)

Biggs, S.D. and Farrington, J. (1991) 'Agricultural research and the rural poor: a review of social science analysis', *Report IDRC - 280e*, Ottawa: International Development Research Centre.

Black, J.K. (1991) *Development in Theory and Practice: Bridging the Gap*, Boulder, Colorado: Westview Press.

Blauert J. (1990) Autochthonous approaches to rural environmental problems: the Mixteca Alta, Oaxaca, Mexico, PhD dissertation, University of London (Wye College).

Bobbio, N. (1987) *The Future of Democracy.* Cambridge: Polity Press.

Bojanic, A. (1991) 'La transferencia de tecnología en Bolivia: la marcha para llegar al modelo de usuarios intermediarios', paper presented at 'Taller Regional para América del Sur: Generación y Transferencia de Tecnología Agropecuaria; el Papel de las ONGs y el Sector Público'. 2–7 December 1991, Santa Cruz, Bolivia.

Booth, D. (1992) 'Social development research: An Agenda for the 1990s'. *European Journal of Development Research* 4(1): 1–39.

—— (ed.) (forthcoming) *New Directions in Social Development Research: Relevance, Realism and Choice.*

Bratton, M. (1989) 'The politics of NGO–Government relations in Africa'. *World Development* 17(4): 569–87.

Brown, L.R. (1970) *Seeds of Change: The Green Revolution and Development in the 1970s*, London: Pall Mall, for Overseas Development Council.

Bruns, B. and Soelaiman, I. (1993) 'Indonesia's Institute for Social and Economic Research, Education and Information: (LP3ES)'s work with small-scale irrigation systems', in J. Farrington and D. Lewis (eds) *NGOs and the State in Asia: Rethinking Roles in Sustainable Agricultural Development*, London: Routledge.

Buck, L. (1993) 'NGOs, government and agroforestry research methodology in Kenya', in K. Wellard, and J. G. Copestake (eds) *NGOs and the State in Africa: Rethinking Roles in Sustainable Agricultural Development*, London: Routledge.

Buckland, J. and Graham, P. (1990) 'The Mennonite Central Committee's experience in agricultural research and extension in Bangladesh', *Agricultural Research and Extension Network Paper No 17*, London: Overseas Development Institute.

Byerlee, D. (1987) *Maintaining the Momentum in Post-Green Revolution Agriculture: A Micro-Level Perspective from Asia*, East Lansing: Michigan State University International Development Paper # 10.

CAAP (1991) 'Generación y Transferencia de Tecnología Agropecuaria. Sistematización de Experiencias en el CAAP', paper presented at 'Taller Regional para América del Sur: Generación y Transferencia de Tecnología Agropecuaria; el Papel de las ONGs y el Sector Público' 2–7 December 1991, Santa Cruz, Bolivia. (Summarized in English as 'El Centro Andino de Acción Popular' in A.J. Bebbington and G. Thiele (eds) *NGOs and the State in Latin America: Rethinking Roles in Sustainable Agricultural Development*, London: Routledge.

Cardoso, V.H. (1991) 'Relación del programa de investigación en producción del INIAP con las organizaciones no gubernamentales y organizaciones campesinas', paper presented at 'Taller Regional para América del Sur: Generación y Transferencia de Tecnología Agropecuaria; el Papel de las ONGs y el Sector Público'. 2–7 December 1991, Santa Cruz, Bolivia.

Carew-Reid, J. and Oli, K.P. (1993) 'The International Union for the Conservation of Nature (IUCN) and NGOs in environmental planning in Nepal', in J. Farrington and D. Lewis (eds) *NGOs and the State in Asia: Rethinking Roles in Sustainable Agricultural Development*, London: Routledge.

Carroll, T. (1964, 1970) 'Land reform as an explosive force in Latin America', in R. Stavenhagen (ed.) *Agrarian problems and Peasant Movements in Latin America*, New York: Anchor Books, pp. 101–38.

—— (1992) *Intermediary NGOs: The Supporting Link in Grassroots Development*, West Hartford, Connecticut: Kumarian Press.

REFERENCES

Carroll, T. Humphreys, D. and Scurrah, M. (1991) 'Grassroots support organizations in Peru', *Development in Practice* 1(2): 97–108.

Cerna, L. and Miclat-Teves, A.G. (1993) 'Mag'uugmad Foundation's (MFI) experience of upland technology development in the Philippines: Soil and water conservation strategies', in J. Farrington and D. Lewis (eds) *NGOs and the State in Asia: Rethinking Roles in Sustainable Agricultural Development*, London: Routledge.

Cernea, M. (1988) *Nongovernmental Organizations and Local Development*, World Bank Discussion Paper Number 40, Washington: World Bank.

CESA (1991) 'La Relación de CESA con el Estado en la Generación y Transferencia de la Tecnología Agropecuaria', paper presented at 'Taller Regional para América del Sur: Generación y Transferencia de Tecnología Agropecuaria; el Papel de las ONGs y el Sector Público'. 2–7 December 1991, Santa Cruz, Bolivia.

Chaguma, A. and Gumbo, D. (1993) 'ENDA-Zimbabwe and Community Research', in K. Wellard, and J. G. Copestake (eds) *NGOs and the State in Africa: Rethinking Roles in Sustainable Agricultural Development*, London: Routledge.

Chakraborty, S., Mandal, B., Das, C., and Satish, S. (1993) 'Ramakrishna Mission: research, extension and training in a farming systems context', in J. Farrington and D. Lewis (eds) *NGOs and the State in Asia: Rethinking Roles in Sustainable Agricultural Development*, London: Routledge.

Chambers, R. and Ghildyal, B.P. (1985) 'Agricultural research for resource-poor farmers: the farmer-first-and-last model', *Agricultural Administration* 20: 1–30.

Chambers, R. and Jiggins, J. (1986) 'Agricultural research for resource-poor farmers: a parsimonious paradigm', *Discussion Paper* No. 220, Brighton: Institute of Development Studies.

Chambers, R. (1983) *Rural Development: Putting the Last First*, Harlow: Longman.

—— (1987) *Sustainable Livelihoods, Environment and Development: Putting Poor Rural People First*, Discussion Paper No. 240, Brighton: Institute of Development Studies.

Chambers, R. (1992) 'Spreading and self-improving: a strategy for scaling-up', in M. Edwards and M.D. Hulme (eds) *Making a Difference in a Changing World*, London: Earthscan Publications.

Chambers, R., Pacey, A., and Thrupp, L.A. (eds) (1989) *Farmer First: Farmer Innovation and Agricultural Research*, London: Intermediate Technology Publications.

Charles, R.A. and Wellard, K. (1993) 'Agricultural activities of government and NGOs in Siaya District' in J. Farrington and D. Lewis (eds) *NGOs and the State in Asia: Rethinking Roles in Sustainable Agricultural Development*, London: Routledge.

Chavez, J. (1991) 'Los programas de agricultura ecológica y de agro-industria alimentaria del centro IDEAS', paper presented at 'Taller Regional para América del Sur: Generación y Transferencia de Tecnología Agropecuaria: el Papel de las ONGs y el Sector Público', 2–7 December 1991, Santa Cruz, Bolivia.

Clark, J. (1991) *Democratizing Development. The Role of Voluntary Organizations*, London: Earthscan Publications.

Colclough, C. and Manor, J. (eds) (1991) *States or Markets: Neoliberalism and the Development Policy Debate*, Oxford: Oxford University Press.

Collinson, M.P. (1972) *Farm Management in Peasant Agriculture: a Handbook for Rural Development Planning in Africa*, New York: Praeger.

Collinson, M. (1988) 'The development of African farming systems: some personal views'. *Agricultural Administration* 29 (1): 7–22.

Copestake, J. (1993) 'The contribution to agricultural technology development of the Gwembe Valley Agricultural Mission (1985–90)', in K. Wellard and J.G. Copestake (eds) *NGOs and the State in Africa: Rethinking Roles in Sustainable Agricultural Development*, London: Routledge.

Corbridge, S. (1982) 'Urban bias, rural bias, and industrialisation: an appraisal of the work of Michael Lipton and Terry Byres', in J. Harriss (ed.) *Rural Development: Theories of Peasant Economy and Agrarian Change*, London: Hutchinson, pp. 94–116.

Cromwell, E. and Wiggins, S. (1993) *Sowing beyond the State: NGOs and Seed Supply in Developing Countries*, London: Overseas Development Institute.

Cotlear, D. 1989. 'The effect of education on farm productivity', *Journal of Development Planning* No 19: 73–99.

Coulter, J. and Farrington, J. (1988) 'India – renewable natural resources sector: review of ODA-supported research and development', Unpublished report for ODA, London.

Dalrymple, D.G. (1986) 'Development and spread of high-yielding varieties of wheat and rice in the less-developed nations', *Foreign Agricultural Economic Report No. 95*, Washington, DC: United States Department of Agriculture.

de Janvry, A. and Garcia, R. (1992) *Rural Poverty and Environmental Degradation in Latin America: Technical Issues in Rural Poverty Alleviation*, Staff Working Paper 1, Rome: International Fund for Agricultural Development.

de Janvry, A., Marsh, R., Runsten, D., Sadoulet, E., and Zabin, C. (1989) 'Impacto de la crisis en la economía campesina de América Latina y el Caribe', in F. Jordan (ed.) *La Economía Campesina: Crisis, Reactivación, Políticas*, San José: Instituto Interamericano de Cooperación para la Agricultura, pp. 91–205.

de Vries, P. (1992) 'Unruly clients. A study of how bureaucrats try and fail to transform gatekeepers, communists and preachers into ideal beneficiaries', PhD thesis, Department of Rural Sociology, Agricultural University, Wageningen, Netherlands.

de Soto, H. (1987) *El Otro Sendero: La Revolución Informal*. Lima, Peru: Instituto Libertad y Democracia.

Dean, L. (1990) 'Local Initiatives for Farmers' Training (LIFT) Gaibandha July 1986–June 1990: Final Report'. CARE Bangladesh, Dhaka (unpublished).

Diop, A.M. (1993) 'Rodale Institute/Rodale International/CRAR (Senegal)' in K. Wellard, and J. G. Copestake (eds) *NGOs and the State in Africa: Rethinking Roles in Sustainable Agricultural Development*, London: Routledge.

Drabek, A. (ed.) (1987) 'Development alternatives: the challenge for NGOs', *World Development* supplement to vol. 15.

Dugue, P. (1993) 'The Senegalese Institute for Research (ISRA) and the Fatick Region Farmers' Association', in K. Wellard, and J. G. Copestake (eds) *NGOs and the State in Africa: Rethinking Roles in Sustainable Agricultural Development*, London: Routledge.

Durán B.J. (1990) *Las Nuevas Instituciones de la Sociedad Civil. Impacto y Tendencia de la Co-operación Internacional y las ONGs en el Area Rural de Bolivia*, La Paz, Bolivia: Huellas.

Edwards, M. (1989) 'The irrelevance of development studies', *Third World Quarterly* 11(1), 116–36.

Edwards, M. and Hulme, D. (eds) (1992a) *Making a Difference: NGOs and Development in a Changing World*, London: Earthscan Publications.

Edwards, M. and Hulme, D. (1992b) 'Scaling up NGO impact on development: learning from experience', *Development in Practice* 2 (2): 77–91.

Ekins, P. (1992) *A New World Order. Grassroots Movements for Global Change*, London: Routledge.

Ekpere, J. and Idowu, I. (1989) *Managing the Links between Research and Technology*

200

Transfer: the Case of the Agricultural Extension-Research Liaison Service in Nigeria, International Service for National Agricultural Research: The Hague.

Esman, M.J. and Uphoff, N. (1984) *Local Organizations: Intermediaries in rural development,* Ithaca, New York: Cornell University Press.

Evans, S. and Boyte, H. (1986) *Free Spaces,* New York: Harper and Row.

Evers, T. (1985) 'Identity: the hidden side of new social movements in Latin America', in D. Slater (ed.) *New Social Movements and the State in Latin America,* Amsterdam: CEDLA.

EXTIE-World Bank (1990) 'Strengthening the Bank's work on popular participation. A proposed learning process', mimeo, Washington: EXTIE-World Bank.

Eyzaguirre, P.B. (1991) 'The scale and scope of national agricultural research in small developing countries: concepts and methodology', *ISNAR Small-Countries Study Paper No.1,* The Hague: International Service for National Agricultural Research.

Fairhead, J. (1990) 'Fields of struggle: towards a social history of farming, knowledge and practice in a Bwisha community, Kivu, Zaire', PhD thesis, Department of Anthropology, School of Oriental and African Studies, University of London.

FAO (1988) *Potentials for Agricultural and Rural Development in Latin America and the Caribbean. Annex V: Crops, Livestock Fisheries and Forestry,* Rome: Food and Agricultural Organization.

Farmer, B.H. (1977) *Green Revolution? Technology and change in rice-growing areas of Tamil Nadu and Sri Lanka,* London: Macmillan.

Farnworth, E.G. (1991) 'The Inter-American Development Bank's interactions with non-governmental environmental organizations', paper prepared for the Third Consultative Meeting on the Environment, 17–19 June 1991, Caracas, Venezuela.

Farrington, J. and Lewis, D. (eds) (1993) *NGOs and the State in Asia: Rethinking Roles in Sustainable Agricultural Development,* London: Routledge.

Farrington, J. and Martin, A. (1988) *Farmer Participation in Agricultural Research: A review of Concepts and Practices.* Occasional Paper No. 9. London. Overseas Development Institute.

Farrington, J. and Mathema, S.B. (1991) 'Managing agricultural research for fragile environments: Amazon and Himalayan case studies', *Occasional Paper No. 11,* London: Overseas Development Institute.

Fernandez, C. and del Rosario, T. (1993) 'The Philippine Department of Agriculture's NGO Outreach Desk', in J. Farrington and D. Lewis (eds) *NGOs and the State in Asia: Rethinking Roles in Sustainable Agricultural Development,* London: Routledge.

Fernandez, A.P. and Mascarenhas, J. (1993) 'MYRADA – participatory rural appraisal and participatory learning methods', in J. Farrington and D. Lewis (eds) *NGOs and the State in Asia: Rethinking Roles in Sustainable Agricultural Development,* London: Routledge.

Figueroa, A. and Bolliger, F. (1986). *Productividad y Aprendizaje en el Medio Ambiente Rural. Informe Comparativo,* Rio de Janeiro: ECIEL.

Fisher-Peck, S. (1993) *The Road from Rio: Sustainable Development and the Non-Governmental Movement in the Third World,* New York, Praeger.

Fowler, A. (1988) 'NGOs in Africa: achieving comparative advantage in relief and micro-development', *Discussion Paper* No. 249. Brighton: Institute of Development Studies.

—— (1990) *Doing it Better? Where and How NGOs have a Comparative Advantage in Facilitating Development,* University of Reading: AERDD Bulletin 28, February.

—— (1991) 'The role of NGOs in changing state-society relations', *Development Policy Review* 9(1): 53–84.

Fox, J. (1990a) 'Editor's introduction', in J. Fox (ed.) *The Challenge of Rural*

Democratisation: Perspectives from Latin America and the Philippines, London: Frank Cass, pp. 1–18.

—— (ed.) (1990b) *The Challenge of Rural Democratisation: Perspectives from Latin America and the Philippines*. London: Frank Cass. (This is also published as a special issue of *Journal of Development Studies* 26 (4).)

Freire, P. (1972) *Pedagogy of the Oppressed*, Harmondsworth: Penguin.

Friedmann, J. (1992) *Empowerment: The Politics of Alternative Development*, Oxford: Blackwell.

Galt, D. and Mathema, S.B. (1987) 'The Samuhik Bhraman process in Nepal: a multidisciplinary group activity to approach farmers'. Paper presented to the Workshop 'Farmers and agricultural research: complementary methods'. Institute of Development Studies, Brighton, 26–31 July.

Garcia, R., Warmenbol, K. and Matsuzaki, S. (1991) 'Transferencia de Tecnología y Desarrollo Comunitario: Experiencia de CIPCA en el Departamento de Santa Cruz, Bolivia', paper presented at 'Taller Regional para América del Sur: Generación y Transferencia de Tecnología Agropecuaria; el Papel de las ONGs y el Sector Público'. 2–7th December 1991, Santa Cruz, Bolivia.

Garforth, C.J. and Munro, M. (1990) *Rural people's organisations and agricultural development in the Upper North of Thailand*, University of Reading, Department of Agricultural Extension and Rural Development.

Giddens, A. (1979) *Central Problems in Social Theory*, London: Macmillan.

Gilbert, E. and Matlon, P. (1992) 'The small W. African NARS – an endangered species?' draft paper for the ISNAR Workshop on Small National Agricultural Research Systems at the University of Mauritius, 28 April – 1 May.

Giordano, E., Satish, S. and Farrington, J. (1993) 'Greenwork at Auroville: from survival to inter-institutional collaboration', in J. Farrington and D. Lewis (eds) *NGOs and the State in Asia: Rethinking Roles in Sustainable Agricultural Development*, London: Routledge.

Gomez, S. and Echenique, J. (1988) *La Agricultura Chilena: las dos Caras de la Modernización*, Santiago, Chile: FLACSO.

Gonsalves, J. and Miclat-Teves, A.G. (eds) (1991) *GO-NGO Collaboration in the Area of Agriculture and Natural Resources Management in the Philippines*, Proceedings of Workshop held by International Institute of Rural Reconstruction (IIRR) and Overseas Development Institute (ODI) at Silang, Cavite, 18–20 July.

Gonsalves, J. and Miclat-Teves, A. (1993) 'The International Institute for Rural Reconstruction (IIRR): developing an agro-forestry kit', in J. Farrington and D. Lewis (eds) *NGOs and the State in Asia: Rethinking Roles in Sustainable Agricultural Development*, London: Routledge.

Gonzalez, W. (1991) 'PROCADE: Un Análisis de Su Rol Coordinador de Actividades Interinstitucionales y de Relación con el Sector Público Agropecuario', paper presented at 'Taller Regional para América del Sur: Generación y Transferencia de Tecnología Agropecuaria; el Papel de las ONGs y el Sector Público', 2–7 December 1991, Santa Cruz, Bolivia.

Goodman, D. and Redclift, M. (eds) (1991) *Environment and Development in Latin America: the Politics of Sustainability*, Manchester: Manchester University Press.

Graham, P. (1992) 'Research and extension of animal draught technologies: the experience of Mbeya Oxenization Project 1987–1992', paper presented at the Conference on Agricultural Research, Training and Technology Transfer in the Southern Highlands of Tanzania: Past Achievements and Future Prospects, Uyole Agriculture Centre, 5–9 October 1992.

Griffin, K. (1975) *The Political Economy of Agrarian Change: An Essay on the Green Revolution*, London: Macmillan.

Grindle, M S. (1986) *State and Countryside. Development Policy and Agrarian Politics in Latin America*, Baltimore: Johns Hopkins University Press.

Gubbels, P. (1992) 'Farmer first research: populist pipedream or practical paradigm? prospects for indigenous agricultural development in West Africa', MA Thesis, School of Development Studies, University of East Anglia, Norwich.

Guerrero, L. (1991) 'La generación y transferencia de tecnologia entre ONG-Taller-Universidad en el Departamento de Cajamarca, Peru', paper presented to the workshop 'Generación y Transferencia de Tecnología agropecuaria: el Papel de las ONG y el Sector Público', 2-7 December 1991, Santa Cruz, Bolivia.

Hashemi, S. M. (1989) 'NGOs in Bangladesh: development alternative or alternative rhetoric?' mimeo, Dhaka.

Harvey, N. (1991) *The New Agrarian Movement in Mexico, 1979-1990*, Research Paper 23, London: Institute of Latin American Studies.

Hassanullah, M. (1991) 'NGOs and public sector agricultural research and extension in Bangladesh', paper presented at the Asia Regional Workshop on NGOs, Natural Resource Management and Links with the Public Sector, Hyderabad, India, 16-20 September 1991.

Healey, J. and Robinson, M. (1992) *Democracy, Governance and Economic Policy: Sub-Saharan Africa in Comparative Perspective*, London: Overseas Development Institute.

Henderson, P.A. and Singh, R. (1990) 'NGO-government links in seed production: case studies from The Gambia and Ethiopia', *Agricultural Research and Extension Network Paper No.14*, London: Overseas Development Institute.

Hewitt de Alcántara, C. (1976) *Modernizing Mexican Agriculture*, Geneva: United Nations Research institute for Social Development.

Hirschmann, A. (1984) *Getting Ahead Collectively: Grassroots Experiences in Latin America*, Oxford: Pergamon.

Hunter, G. (1982) 'Enlisting the small farmer: the range of requirements', Agricultural Administration Unit Occasional Paper No. 4, London: Overseas Development Institute.

Hyden, G. (1983) *No Shortcuts to Progress. African Development Management in Perspective*, Berkeley: University of California Press.

Hyman, E.L. (1992) 'Local agroprocessing with sustainable technology: sunflower seed oil in Tanzania', *Gatekeeper Series No.33*, London: International Institute for Environment and Development.

IFAD (1992) *The State of World Poverty. An Inquiry into its Causes and its Consequences*, New York: published for the International Fund for Agricultural Development by New York University Press.

IICA (1991) *Regional Overview of Food Security in Latin America and the Caribbean with a Focus on Agricultural Research and Technology Transfer*, Programa de Generación y Transferencia de Tecnología, San José, Costa Rica: Instituto Interamericano de Cooperación para la Agricultura.

ISNAR (1992) *Summary of Agricultural Research Policy: International Quantitative Perspectives*, The Hague: International Service for National Agricultural Research.

Jodha, N.S. (1986) 'Common property resources and the rural poor in dry regions of India', *Economic and Political Weekly* 21(27): 1169-81.

Jonjuabsong, L. and Hwai-Kham, A. (1993) 'The Appropriate Technology Association: promoting rice-fish farming and biological pesticides', in J. Farrington and D. Lewis (eds) *NGOs and the State in Asia: Rethinking Roles in Sustainable Agricultural Development*, London: Routledge.

Kabeer, N. (1989) 'Monitoring poverty as if gender mattered' Discussion Paper No. 255, Brighton: Institute of Development Studies.

Kaimowitz, D. (ed.) (1990) *Making the Link: Agricultural Research and Technology Transfer Services in Developing Countries*, Boulder, Colorado, and London: Westview Press (in collaboration with ISNAR).

Kaimowitz, D. (1991) 'El papel de las ONG en el sistema Latinoamericano de generación y transferencia de tecnología agropecuaria', paper presented to the workshop 'Generación y Transferencia de Tecnología Agropecuaria: el Papel de las ONG y el Sector Público', 2–7 December 1991, Santa Cruz, Bolivia.

Kaimowitz, D., Snyder, M., and Engel, P. (1990) 'A conceptual framework for studying the links between agricultural research and technology transfer in developing countries', in D. Kaimowitz (ed.) *Making the Link: Agricultural Research and Technology Transfer Services in Developing Countries*, Boulder, Colorado and London: Westview Press (in collaboration with ISNAR), pp. 227–69.

Kashem, M., Mozahar Ali, M. Halim, A. Monirul Islam, M. Uddin Milki, M.G. Hannan Bhuiya, A., and Islam Bhuiya, S. (eds) (1991) 'Proceedings of the Second National Seminar on GO–NGO Collaboration in Agricultural Research and Extension, 4 August 1991', *Bangladesh Journal of Extension Education* vol. 6, Special issue.

Kates, K.W. and Haarman, V. (1992) 'Where the poor live: are the assumptions correct?' *Environment* 34(4) pp. 4–11, 25–8.

Kean, S. and Singogo, L. (1990) *Bridging the Gap between Research and Extension in Zambia: the Incorporation of Research-extension Liaison Officers into the Adaptive Research Planning Team*, The Hague: ISNAR.

Kent, R.C. (1987) *Anatomy of Disaster Relief*, London: Pinter.

Khan, S.S. (1992) 'NGO alternatives and fresh initiatives in extension – the Aga Khan Rural Support Programme Experience', paper presented to the 12th Agricultural Symposium, 8–10 January 1992, held at the World Bank, Washington DC.

Khan, M., Lewis, D.J., Sabri, A. A., and Shahabuddin, M. (1993) 'NGO work in livestock and social forestry technologies in Bangladesh – the case of Proshika', in J. Farrington and D. Lewis (eds) *NGOs and the State in Asia: Rethinking Roles in Sustainable Agricultural Development*, London: Routledge.

Kleemeyer, C. (1991) 'What is grassroots development?' *Grassroots Development* 15(1): 38–39.

Kleemeyer, C. (1992) 'Cultural energy and grassroots development', *Grassroots Development* 16(1) 22–31.

Kohl, B. (1991) 'Protected horticultural systems in the Bolivian Andes: a case study of NGOs and inappropriate technology'. *Agricultural Research and Extension Network Paper No. 29*, London: Overseas Development Institute.

Kolbilla, D. and Wellard, K. (1993) 'Langbensi Agricultural Station: Experiences of Agricultural Research' in K. Wellard and J.G. Copestake (eds) *NGOs and the State in Africa: Rethinking Roles in Sustainable Agricultural Development*, London: Routledge.

Kopp, A. and Domingo, T. (1991) 'Tecnologías de conservación en el trópico: CESA, Bolivia', paper presented to the workshop 'Generación y Transferencia de Tecnología Agropecuaria: el Papel de las ONG y el Sector Público,' 2–7 December 1991, Santa Cruz, Bolivia.

Korten, D.C. (1987) 'Third generation NGO strategies: a key to people-centred development', *World Development* 15, supplement: 145–59.

—— (1990) *Getting to the 21st Century: Voluntary Action and the Global Agenda*, West Hartford, Connecticut: Kumarian Press.

Korten, F. and Siy, R.F. (1988) *Transforming a bureaucracy*, West Hartford, Connecticut: Kumarian Press.

Krueger, A. (1974) 'Political economy of the rent seeking society', *American Economic Review*, 64(3), 291–303.

Krueger, A., Schiff, M., and Valdes, A. (1991) *The Political Economy of Agricultural Pricing Policy, Volume 1: Latin America*, Baltimore: Johns Hopkins University Press.

Laclau, E. (1985) 'New social movements and the plurality of the social', in D. Slater (ed.) *New Social Movements and the State in Latin America*, Amsterdam: CEDLA, pp. 27–44.

Landsberger, H. and Hewitt, C. (1970) 'Ten sources of weakness and cleavage in Latin American peasant movements', in R. Stavenhagen (ed.) *Agrarian Problems and Peasant Movements in Latin America*, Garden City, NY: Anchor Books, pp. 559–83.

Leach, M. and Mearns, R. (1992) *Poverty and Environment in Developing Countries. An Overview Study*, Report to Economic and Social Research Council, Brighton: Institute of Development Studies.

Lehman, H.P. (1990) 'The politics of adjustment in Kenya and Zimbabwe: the state as intermediary', *Studies in Comparative International Development* 25 (3): 37–72.

Lehmann, A.D. (1982) 'Beyond Lenin and Chayanov. New paths of agrarian capitalism', *Journal of Development Economics* 11: 133–61.

—— (1984) *Economic Development and Social Differentiation in the Andean Peasant Economy*, Studies in Peasant Economy, Cambridge: University of Cambridge Department of Applied Economics.

—— (1990) *Democracy and Development in Latin America. Economics, Politics and Religion in the Postwar Period*, Cambridge: Polity Press.

Leonard, H.J. (ed.) (1989) *Environment and the Poor: Development Strategies for a Common Agenda*, Washington DC: Overseas Development Council.

Lewis, D.J. (1993) *Going it Alone: A Literature Review of Female-headed Households in Bangladesh*, Occasional Paper, Centre for Development Studies, University of Bath.

Lipton, M. (1977; 1986) *Why Poor People Stay Poor: A Study of Urban Bias in World Development*, 1st edition – London: Temple Smith; 2nd edition – Aldershot: Gower.

—— (1988) The place of agricultural research in the development of sub-Saharan Africa, *World Development* 16(10): 1231–57.

Lipton, M. and Longhurst, R. (1989) *New Seeds and Poor People*, London: Unwin Hyman.

Long, N. (1988) 'Sociological perspectives on agrarian development and state intervention', in A. Hall and J. Midgley (eds) *Development Policies: Sociological Perspectives*', Manchester: Manchester University Press.

Long, N. and van der Ploeg, J. (forthcoming) 'Reflections on the actor-oriented approach to social development research: towards a new concept of structure', in D. Booth (ed.) *New Directions in Social Development Research: Relevance, Realism and Choice*.

Longhurst, R. (ed.) (1981) 'Rapid rural appraisal: social structure and rural economy', *Bulletin of the Institute of Development Studies* 12(4), special issue.

Lovell, C. (1992) *Breaking the Cycle of Poverty: The BRAC Strategy*, West Hartford, Connecticut: Kumarian Press.

Loveman, B. (1991) 'NGOs and the transition to democracy in Chile,' *Grassroots Development* 15(2): 8–19.

McGarry, B. (1993) 'Silveira House: propagation of the use of hybrid seed (1968–83)' in K. Wellard, and J.G. Copestake (eds) *NGOs and the State in Africa: Rethinking Roles in Sustainable Agricultural Development*, London: Routledge.

Martínez Noguiera, R. (1990) 'The effect of changes in state policy and organizations on agricultural research and extension links: a Latin American perspective', in D. Kaimowitz (ed.) *Making the Link: Agricultural Research and Technology Transfer*

Services in Developing Countries, Boulder, Colorado and London: Westview Press (in collaboration with ISNAR), pp. 75–108.

Mascarenhas, J., Shah, P., Joseph, S., Jayakaran, R., Devavaram, J., Ramachandran, V., Fernandez, A., Chambers, R., and Pretty, J. (eds) (1991) 'Participatory rural appraisal': proceedings of the February 1991 Bangalore PRA Trainers Workshop, *RRA Notes Number 13, August*, London: IIED/Bangalore: MYRADA.

Mellor, J.W. (1974) 'Economic and social implications and choices related to change in agricultural technology'. Paper 21 in H. Bunting (ed.) *Proceedings of the Second International Seminar on Change in Agriculture*, 9–19 September, University of Reading.

—— (1988) 'Agricultural development opportunities for the 1990s – the role of research', address to the International Centers' Week of the Consultative Group on International Agricultural Research, Washington DC, 4 November.

Merrill-Sands, D. and Kaimowitz, D. (1991). *The Technology Triangle: Linking Farmers, Technology Transfer Agents and Agricultural Researchers*, The Hague: International Service for National Agricultural Research.

Miclat-Teves, A.G. and Lewis, D.J. (1993) 'Philippines overview', in J. Farrington and D. Lewis (eds) *NGOs and the State in Asia: Rethinking Roles in Sustainable Agricultural Development*, London: Routledge.

Millar, D. (1991) 'Networking between Government and Non-Governmental Organisations in Northern Ghana', mimeo, Tamale: Catholic Archdiocese.

Montgomery, J.D. (1988) *Bureaucrats and People: Grassroots Particiapation in Third World Development*, Baltimore and London: Johns Hopkins University Press.

Morgan, M. (1990) 'Stretching the development dollar: the potential for scaling-up', *Grassroots Development* 14(1): 2–12.

Moris, J. (1991) 'Extension alternatives in tropical agriculture', *Occasional Paper No.7*, London: Overseas Development Institute.

Moseley, P. Harrigan, J., and Toye, J. (1991) *Aid and Power: The World Bank amd Policy Based Lending, Volume 1*, London: Routledge.

Munck, G. (1992) Review article, *Journal of Development Studies* 29(1): 176–81.

Mung'ala, P. (1993) 'Government experiences of collaboration with NGOs in rural afforestation', in K. Wellard, and J.G. Copestake (eds) *NGOs and the State in Africa: Rethinking Roles in Sustainable Agricultural Development*, London: Routledge.

Murray, G.F. (1986) 'Seeing the forest while planting the trees: an anthropological approach to agroforestry in rural Haiti', in D.W. Brinkerhoff and J.-C. Garcia-Zamor (eds) *Politics, Projects and People: Institutional Development in Haiti*, New York: Praeger.

Mustafa, S., Rahman, S., and Sattar, G. (1993) 'Bangladesh Rural Advancement Committee (BRAC) – backyard poultry and landless irrigators programmes' in J. Farrington and D. Lewis (eds) *NGOs and the State in Asia: Rethinking Roles in Sustainable Agricultural Development*, London: Routledge.

Mvududu, S. (1993) 'Forestry Commission Links with NGOs in Rural Afforestation' in in K. Wellard, and J.G. Copestake (eds) *NGOs and the State in Africa: Rethinking Roles in Sustainable Agricultural Development*, London: Routledge.

Nahas, F. (1993) 'Appropriate technologies for improved duck-rearing in Bangladesh – the experience of Friends in Village Development', in J. Farrington and D. Lewis (eds) *NGOs and the State in Asia: Rethinking Roles in Sustainable Agricultural Development*, London: Routledge.

Ndiweni, M. (1993) 'The organization of rural associations for progress and grassroots development', in in K. Wellard, and J.G. Copestake (eds) *NGOs and the State in Africa: Rethinking Roles in Sustainable Agricultural Development*, London: Routledge.

REFERENCES

Norman, D.W. (1974) 'Rationalising mixed cropping under indigenous conditions: the example of Northern Nigeria', *Journal of Development Studies* 11(1): 3–21.

—— (1982) 'The farming systems approach to research', *FSR Paper Series No.3*, Manhattan, Kansas: Kansas State University.

Norman, D.W. and Modiakgotla, E. (1990) 'Ensuring farmer input into the research process within an institutional setting: the case of semi-arid Botswana', *Agricultural Research and Extension Network Paper No.16*, London: Overseas Development Institute.

Nunberg, B. (1988) *Public Sector Management Issues in Structural Adjustment Lending*, World Bank Discussion Paper 99, Washington DC: World Bank Publications.

Orr, A., Nazrul Eskim, A.S.M. and Barnes, G. (1991) *The Treadle Pump: Manual Irrigation for Small Farmers in Bangladesh*, Dhaka: PACT.

Otieno, P. (1992) 'Experiences of CARE-Kenya in collaborating with other agencies in Agroforestry in western Kenya', in K. Wellard, G. Arum, and K. Kiambi (eds) *Inter-agency Collaboration in Agricultural and Environmental Technologies*, Proceedings of the Workshop held at Masinga Tourist Resort, Kenya. London: Overseas Development Institute and Nairobi: KENGO.

Owens, S. (1993) 'Catholic relief services in The Gambia: evolution from agricultural research to community-based experimentation', in in K. Wellard, and J.G. Copestake (eds) *NGOs and the State in Africa: Rethinking Roles in Sustainable Agricultural Development*, London: Routledge.

Overseas Development Institute (1992) *Aid and Political Reform*, Briefing Paper, January 1991, London: Overseas Development Institute.

Pardey, P.G., Roseboom, J., and Anderson, J.R. (eds) (1991) *Agricultural Research Policy: International Quantitative Perspectives*, Cambridge: Cambridge University Press.

Paul, S. and Israel, A. (eds) (1991) 'Non-governmental organizations and the World Bank: an overview', in: *Non-Governmental Organisations and the World Bank: Cooperation for Development*, Washington, DC: World Bank.

Pearse, A. (1980) *Seeds of Plenty, Seeds of Want: Social and Economic Implications of the Green Revolution*, Oxford: Oxford University Press.

Poffenberger, M. (ed.) (1990) 'Forest management partnerships: regenerating India's forests', executive summary of the workshop on sustainable forestry, New Delhi, 10–12 September 1990, Delhi: Ford Foundation and Indian Environmental Society.

Pray, C. and Echeverría, R. (1990) 'Private sector agricultural research and technology transfer links in developing countries', *Linkages Theme Paper No. 3*. The Hague: International Service for National Agricultural Research.

Pretty, J. and Chambers, R. (1992) 'Turning the new leaf: new professionalism, institutions and policies for agriculture', paper prepared for 'Beyond Farmer First' Workshop, Institute of Development Studies, University of Sussex, Brighton, 27–29 October 1992.

Redclift, M. (1988) 'Introduction: agrarian social movements in contemporary Mexico', *Bulletin of Latin American Research* 7(2): 249–56.

Rhoades, R.E. and Booth, R. (1982) 'Farmer-back-to-farmer. A model for generating acceptable agricultural technology', *Agricultural Administration* 11: 127–37.

Richards, P. (1983) 'Ecological change and the politics of African land use', *African Studies Review* 26: 1–72.

Richards, P. (1985) *Indigenous Agricultural Revolution: Ecology and Food Production in West Africa*, London: Hutchinson.

Riddell, R. and Robinson, M. (1993) '*Working with the Poor: NGOs and Rural Poverty Alleviation*' manuscript, London: Overseas Development Institute.

Rigg, J. (1989) 'The new rice technology and agrarian change: guilt by association?', *Progress in Human Geography* 13(2): 374–99.

Ritchey-Vance, M. (1991) *The Art of Association: NGOs and Civil Society in Colombia*, Washington, DC: Inter-American Foundation.

Rivera-Cusicanqui, S. (1990) 'Liberal democracy and Ayllu democracy in Bolivia: the case of Northern Potosi,' in J. Fox (ed.) *The Challenge of Rural Democratisation: Perspectives from Latin America and the Philippines*, London: Frank Cass, pp. 97–121.

Robinson, M.A. (1991) 'Evaluating the impact of NGOs in rural poverty alleviation – India country study', *Working Paper No.49*, London: Overseas Development Institute.

Robinson, M.A., Farrington, J., and Satish, S. (1993) 'India overview' in J. Farrington and D. Lewis (eds) *NGOs and the State in Asia: Rethinking Roles in Sustainable Agricultural Development*, London: Routledge.

Ruttan, V. (1991) 'Challenges to agricultural research in the 21st century', in P.G. Pardey, J. Roseboom, and J.R. Anderson (eds) *Agricultural Research Policy: International Quantitative Perspectives*, Cambridge: Cambridge University Press, pp. 399–411.

Sarch, M.-T. and Copestake, J. (1993) 'The Gambia: country overview' in K. Wellard, and J.G. Copestake (eds) *NGOs and the State in Africa: Rethinking Roles in Sustainable Agricultural Development*, London: Routledge.

Sarch, M.-T. (1993) 'Case study of the Farmer Innovation and Technology Testing Programme in The Gambia' in K. Wellard, and J.G. Copestake (eds) *NGOs and the State in Africa: Rethinking Roles in Sustainable Agricultural Development*, London: Routledge.

Satish, S. and Farrington, J. (1990) 'A research-based NGO in India: the Bharatiya Agro-Industries Foundation's Cross-bred Dairy Programme', *Agricultural Research and Extension Network Paper 18*. London: Overseas Development Institute.

—— (1993) 'Bharatiya Agro-Industries Foundation (BAIF) – research programmes in livestock production, health and nutrition', in J. Farrington and D. Lewis (eds) *NGOs and the State in Asia: Rethinking Roles in Sustainable Agricultural Development*, London: Routledge.

Satish, S. and Kamal Kar (1993) 'NGO links with government farm science centres (KVKs) – Seva Bharati: KVK', in J. Farrington and D. Lewis (eds) *NGOs and the State in Asia: Rethinking Roles in Sustainable Agricultural Development*, London: Routledge.

Satish, S. and Kumar, P. (1993) 'Are NGOs more cost-effective than government in livestock service delivery? A study of artificial insemination in India', in J. Farrington and D. Lewis (eds) *NGOs and the State in Asia: Rethinking Roles in Sustainable Agricultural Development*, London: Routledge.

Satish, S. and Dipankar Saha (1993) 'NGO links with government farm science centres – Nimpith KVK' in J. Farrington and D. Lewis (eds) *NGOs and the State in Asia: Rethinking Roles in Sustainable Agricultural Development*, London: Routledge.

Satish, S. and Vardhan, T.J.P.S. (1993) 'Action for world solidarity and the red-headed hairy caterpillar', in J. Farrington and D. Lewis (eds) *NGOs and the State in Asia: Rethinking Roles in Sustainable Agricultural Development*, London: Routledge.

Sayer, A. (1984) *Method in Social Science: A Realist Approach*, London: Hutchinson.

Schultz, T. (1964) *Transforming Traditional Agriculture*, New Haven, Connecticut: Yale University Press.

Scoones, I. and Thompson, J. (1992) 'Beyond farmer first: rural people's knowledge, agricultural research and extension practice: towards a theoretical framework', paper prepared for 'Beyond Farmer First' Workshop, Institute of Development Studies, University of Sussex, Brighton, 27–29 October 1992.

Sen, A. (1981) *Poverty and Famine: An Essay on Entitlement and Deprivation*, Oxford: Clarendon Press.

Sethna, A. and Shah, A. (1993) 'The Aga Khan Rural Support Project: influencing wasteland development policy', in J. Farrington and D. Lewis (eds) *NGOs and the State in Asia: Rethinking Roles in Sustainable Agricultural Development*, London: Routledge.

Seth, S.L. and Axinn, G.H. (1991) 'Institutionalizing a farming systems approach: a case from India', paper presented to the 11th International Farming Systems Research Symposium, East Lansing, Michigan: Michigan State University.

Shah, P. and Mane, P.M. (1993) 'Networking in participatory training and extension: experiences of the Aga Khan Rural Support Programme in Gujarat State', in J. Farrington and D. Lewis (eds) *NGOs and the State in Asia: Rethinking Roles in Sustainable Agricultural Development*, London: Routledge.

Sharma, A.R., Krishna, K.C., Mathema, S.B., and Sharma Paudyal, B.K. (1988) 'Effectiveness and performance survey report on Surkhet agriculture station', *Report No.13*, Kathmandu: SERED.

Shrestha, N.K. and Farrington, J. (1993) 'Nepal overview', in J. Farrington and D. Lewis (eds) *NGOs and the State in Asia: Rethinking Roles in Sustainable Agricultural Development*, London: Routledge.

Slater, D. (1985a) 'Social movements and a recasting of the political', in D. Slater (ed.) *New Social Movements and the State in Latin America*, Amsterdam: CEDLA, pp. 1–25.

Slater, D. (ed.) (1985b) *New Social Movements and the State in Latin America*, Amsterdam: CEDLA.

Smillie, I. (1991) *Mastering the Machine: Poverty, Aid and Technology*, London: IT Publications.

Soliz, R., Espinosa, P., and Cardoso, V.H. (1989) *Ecuador: Organización y Manejo de la Investigación en Finca en el Instituto Nacional de Investigaciones Agropecuarias (INIAP)*, OFCOR Case Study No. 7, The Hague: International Service for National Agricultural Research.

Sollows, J., Jonjuabsong, L., and Hwai-Kham, A. (1991) 'NGO–government interaction in rice-fish farming and other aspects of sustainable agricultural development in Thailand', *Agricultural Research and Extension Network Paper No.28*, London: Overseas Development Institute.

Sollows, J., Thongpan, N., and Leelapatra, W. (1993) 'NGO-Government interaction in rice-fish culture in NE Thailand' in J. Farrington and D. Lewis (eds) *NGOs and the State in Asia: Rethinking Roles in Sustainable Agricultural Development*, London: Routledge.

Sotomayor, O. (1991) 'GIA and the New Chilean Public Sector: the Dilemmas of Successful NGO Influence over the State', *Agricultural Research and Extension, Network Paper No. 30*. London: Overseas Development Institute.

Sumberg, J.E. (1991) 'NGOs and agriculture at the margin: research, participation and sustainability in West Africa', *Agricultural Research and Extension Network Paper No. 27*, London: Overseas Development Institute.

Swaminathan, P. and Satish, S. (1993) 'NGO links with government farm science centres – UPASI KVK', in J. Farrington and D. Lewis (eds) *NGOs and the State in Asia: Rethinking Roles in Sustainable Agricultural Development*, London: Routledge.

Tandon, R. (1987) 'The relationship between NGOs and government', paper presented to the Conference on the Promotion of Autonomous Development, New Delhi: PRIA.

Taylor, J. (1979) *From Modernisation to Modes of Production.*

Tendler, J. (1982) *Turning Private Voluntary Organizations into Development Agencies: Questions for Evaluation*, Program Evaluation Discussion Paper Number 12. Washington, DC: U.S. Agency for International Development.

—— (1987) 'Whatever happened to poverty alleviation? Report for the mid-decade review of the Ford Foundation's Programs on Livelihood, Employment and Income Generation', mimeo, New York: Ford Foundation.

Tendler, J. with Healy, K., and O'Laughlin, C.M. (1988) *What to Think About Cooperatives: A Guide from Bolivia*, in S. Annis and P. Hakim (eds) *Direct to the Poor: Grassroots Development in Latin America*, Boulder and London: Lynne Reinner, pp.85–116.

Thapa, D. (1993) 'United Mission to Nepal's forestry activities, with special reference to the Nepal Coppice Reforestation Programme', in J. Farrington and D. Lewis (eds) *NGOs and the State in Asia: Rethinking Roles in Sustainable Agricultural Development*, London: Routledge.

Thiele, G., Davies, P. and Farrington, J. (1988) 'Strength in diversity: innovation in agricultural technology development in Eastern Bolivia', *Agricultural Research and Extension Network Paper No.1*, London: Overseas Development Institute.

Thiesenhusen, W. (ed.) (1989) *Searching for Agrarian Reform in Latin America*, London: Unwin Hyman.

Tiffen, M. and Mortimore, M. (1992) 'Environment, population growth and productivity in Kenya: a case study of Machakos District', *Development Policy Review* 10(4): 359–88.

Tomboc, C.C. and Reyes, G.D. (1993) 'The Ecosystems Research and Development Bureau's (ERDB) Integrated Livestock Programme: working with people's organizations in the Philippines', in J. Farrington and D. Lewis (eds) *NGOs and the State in Asia: Rethinking Roles in Sustainable Agricultural Development*, London: Routledge.

Trujillo, G. (1991) 'Investigación y Extensión por El Ceibo en el Alto Beni' (Central Regional Agropecuaria-Industrial de Cooperativas 'El Ceibo' Limitada, El Ceibo – Bolivia), paper presented to the workshop 'Generación y Transferencia de Tecnología Agropecuaria: el Papel de las ONG y el Sector Público,' 2–7 December 1991, Santa Cruz, Bolivia.

Turner, B.L. and Benjamin, P. (1991) 'Fragile Lands: Identification and Use for Agriculture', paper prepared for the Conference on Institutional Innovations for Sustainable Agricultural Development: Into the Twenty-first Century, Bellagio, Italy 14–18 October 1991.

Turner, B.L. and Brush, S. (1987) *Comparative Farming Systems*, New York: Guildford Press.

United Nations, (1988) *World Demographic Estimates and Projections, 1950-2025*, New York: United Nations.

UNDP (1990) *Human Development Report 1990*, Oxford: Oxford University Press.

—— (1992) *Human Development Report 1992*, Oxford: Oxford University Press.

Uphoff, N. (1986) *Local Institutional Development. An Analytical Sourcebook with Cases.* West Hartford, Connecticut: Kumarian Press.

—— (1992) *Learning from Gal Oya: Possibilities for Participatory Development and Post-Newtonian Social Science*, Ithaca, New York and London: Cornell University Press.

—— (1993) 'Grassroots organizations and NGOs in rural development: opportunities with diminishing states and expanding markets', paper for the conference 'State,

Market and Civil Institutions: New Theories, New Practices and their Implications for Rural Development, 13–14 December 1991. Published 1993, in *World Development* 21(4).

Vasimalai, M. (1993) 'Professional Assistance for Development Action (PRADAN): an NGO de-mystifies and scales down technology', in J. Farrington and D. Lewis (eds) *NGOs and the State in Asia: Rethinking Roles in Sustainable Agricultural Development*, London: Routledge.

Velez, R. and Thiele, G. (1991) 'Primeras experiencias con un nuevo modelo de transferencia de tecnología', paper presented to the workshop 'Generación y Transferencia de Tecnología Agropecuaria: el Papel de las ONG y el Sector Público', 2–7 December 1991, Santa Cruz, Bolivia.

Watson, H.R. and Laquihon, W.A. (1993) 'The Mindanao Baptist Rural Life Center's Sloping Agricultural Land Technology (SALT) research and extension in the Philippines', in J. Farrington and D. Lewis (eds) *NGOs and the State in Asia: Rethinking Roles in Sustainable Agricultural Development*, London: Routledge.

Wellard, K. and Copestake, J.G. (eds) (1993) *NGOs and the State in Africa: Rethinking Roles in Sustainable Agricultural Development*, London: Routledge.

Wellard, K. and Mema, N. (eds) (1991) *NGO-Government Collaboration in Agricultural and Environmental Technologies*, Proceedings of the Workshop held in Harare, Zimbabwe, 11–12 March 1991, London: Overseas Development Institute/Harare: NANGO.

Wellard, K., Arum, G., and Kiambi, K. (eds) (1991) *Inter-agency Collaboration in Agricultural and Environmental Technologies*, Proceedings of the workshop held at Masinga Tourist Resort, Kenya, London: Overseas Development Institute and Nairobi: KENGO.

Williams, A. (1990) 'A growing role for NGOs in development', *Finance and Development*, December: 31–3.

Wils, F. (1990) 'Income and employment-generating activities of NGOs: an overview', working paper 88, The Hague: Institute of Social Studies.

Wilson, M. (1991) 'Reducing the costs of public extension services: initiatives in Latin America', in W.M. Rivera and D.J. Gustafson (eds) *Agricultural Extension: Worldwide Evolution and Forces for Change*, Amsterdam: Elsevier.

Wood, G. D. and Palmer-Jones, R. (1990) *The Water Sellers*, London: Intermediate Technology Publications and West Hartford, Connecticut: Kumarian Press.

World Bank (1990a) *How the World Bank Works with Nongovernmental Organizations*. Washington, DC: World Bank.

—— (1990b) *Social Indicators of Development*, Baltimore: Johns Hopkins University Press.

—— (1990c) *Social Investment in Guatemala, El Salvador, and Honduras*, report on the Workshop on Poverty Alleviation, Basic Social Services and Social Investment Funds, Paris, 29–30 June 1990, Washington, DC: World Bank.

—— (1991a) *World Development Report 1991. The Challenge of Development*, Oxford: Oxford University Press.

—— (1991b) *Trends in Developing Economies 1991*, Washington, DC: World Bank.

—— (1992) *Poverty Handbook, Discussion Draft*, Washington DC: World Bank.

—— (n.d.), 'The Bank and NGOs: recent experience and emerging trends', draft. Washington, DC: World Bank.

INDEX

Note: Page numbers in **bold** type indicate boxes giving relevant case study notes. Abbreviations: With the exception of NGO (non-governmental organization) abbreviations are used only in subheadings and as cross references to the main subject headings (i.e. IAF *see* Inter-American Foundation). A full glossary can be found on pages xxiv–xxvii.